Autonomous Underwater Vehicles

Modeling, Control Design, and Simulation

Autonomous Underwater Vehicles

Modeling, Control Design, and Simulation

Sabiha Wadoo • Pushkin Kachroo

CRC Press
Taylor & Francis Group
Boca Raton London New York

CRC Press is an imprint of the
Taylor & Francis Group, an **informa** business

CRC Press
Taylor & Francis Group
6000 Broken Sound Parkway NW, Suite 300
Boca Raton, FL 33487-2742

Printed in the United States of America on acid-free paper
10 9 8 7 6 5 4 3 2 1

International Standard Book Number: 978-1-4398-1831-2 (Hardback)

Library of Congress Cataloging-in-Publication Data

Wadoo, Sabiha.
Autonomous underwater vehicles : modeling, control, design, and simulation / Sabiha Wadoo, Pushkin Kachroo.
p. cm.
Includes bibliographical references and index.
ISBN 978-1-4398-1831-2 (hardcover : alk. paper)
1. Remote submersibles--Mathematical models. 2. Remote submersibles--Design and construction. 3. Autonomous robots--Design and construction. 4. Feedback control systems. 5. Nonlinear control theory. I. Kachroo, Pushkin. II. Title.

TC1662.W33 2011
623.82'05--dc22 2010043678

Visit the Taylor & Francis Web site at
http://www.taylorandfrancis.com

and the CRC Press Web site at
http://www.crcpress.com

Dedication

To my children Saami and Shireen Sabiha Wadoo

To my children Axenya and Sheen Pushkin Kachroo

Contents

Preface

Control design of autonomous underwater vehicles is an important area for researchers in the control systems community. Control is generally difficult to achieve due to nonlinear dynamics, uncertain models, and the presence of disturbances that are hard to measure or estimate. This book presents a new approach to the modeling and control design of autonomous underwater vehicles.

Kinematic and dynamic nonlinear models for autonomous underwater vehicles are discussed in this book. Controllability of autonomous underwater vehicles for different motion planning tasks is reviewed. The book combines the feedback control design for both kinematic and dynamic models. The feedback control is also achieved in the presence of uncertainties, thereby making the control design robust.

This work is intended for students and researchers, as there is more development that is needed in this particular area. The results of this book can be extended to obtain advanced control strategies and design schemes for autonomous underwater vehicles and other similar problems falling in the area of nonlinear control.

This research was partially supported by a grant from New York Institute of Technology's Institutional Support of Research and Creativity (ISRC).

For product information about MATLAB®, please contact:
The MathWorks, Inc.
3 Apple Hill Drive
Natick, MA 01760-2098 USA
Tel: 508-647-7000
Fax: 508-647-7001
E-mail: info@mathworks.com
Web: www.mathworks.com

Sabiha Wadoo
New York Institute of Technology

About the Authors

Sabiha Wadoo, PhD, received a BE degree in electrical engineering from the Regional Engineering College, Kashmir, India, in 2001, and an MS degree in electrical engineering, an MS degree in mathematics, and a PhD degree in electrical engineering from Virginia Polytechnic Institute and State University (Virginia Tech), Blacksburg, in 2003, 2005, and 2007, respectively.

Since 2007, she has been with the New York Institute of Technology, Old Westbury, New York, where she is an assistant professor with the Department of Electrical and Computer Engineering. Her research interests are in the areas of feedback control of nonlinear control systems, nonlinear control system abstraction, and feedback control of distributed parameter systems.

Pushkin Kachroo, PhD, received a BTech degree in civil engineering from the Indian Institute of Technology, Bombay, India, in 1988, an MS degree in mechanical engineering from Rice University, Houston, Texas, in 1990, a PhD degree in mechanical engineering from the University of California, Berkeley, in 1993, and MS and PhD degrees in mathematics from Virginia Polytechnic Institute and State University (Virginia Tech), Blacksburg, in 2004 and 2007, respectively.

He is the director of the Transportation Research Center, Harry Reid Center for Environmental Studies, Las Vegas, Nevada, and a professor with the Department of Electrical and Computer Engineering, University of Nevada, Las Vegas.

1 Introduction

1.1 OVERVIEW

The work presented in this book is concerned with developing models that can be used for the motion planning and feedback control design of underwater vehicles. Underwater vehicles present difficult control system design problems due to their nonlinear dynamics, uncertain models, and the presence of disturbances that are difficult to measure or estimate. Many problems must be solved to make robotic underwater vehicles a reality. The dynamic control of the vehicle needs to guarantee stability and perform consistently. The dynamics of autonomous underwater vehicles present a difficult control system design problem that traditional linear design methodologies cannot accommodate easily. The dynamics are fundamentally nonlinear in nature. Hydrodynamic coefficients often are poorly known, and a variety of immeasurable disturbances are present due to currents.

In this book, a new approach of feedback control methodology is developed to the accurate trajectory control and point stabilization of underwater vehicles. The feedback control is developed in both the absence and presence of uncertainties. The methods can deal with nonlinear dynamics directly, and are shown to be stable in the presence of disturbances, thereby making the control robust.

The first part of the book presents the study concerning the applicability of kinematic-based control of underwater vehicles. The methods of motion planning and feedback control for a kinematic model of an underwater vehicle are developed. The kinematic model of an underwater vehicle falls into a general category of nonholonomic systems. The system is characterized by nonholonomic constraints on its generalized velocities. The motion planning problem for this system, with constraints on the velocities, is transformed into a control problem having fewer control inputs than the degrees of freedom. The nonlinear controllability issues for the system are also studied. For the design of feedback controllers, the system is transformed into chained and power forms. The methods of transforming the kinematic model of the system into these forms are discussed. Differential geometric control theory and nonlinear system analysis and control design techniques, and the results of recent research and study in the motion planning of nonholonomic systems are used and presented for the support of the work.

The kinematic model of the autonomous underwater vehicle is developed, and the feedback controller design for the same is presented in detail and simulation results are obtained. Also, a brief mathematical analysis of the concepts involved in the study of controllability, control design, and modeling is presented.

The kinematic-based control addresses the motion planning for a kinematic model of an underwater vehicle. The kinematic model belongs to nonholonomic systems. The control model for such systems is drift-free, nonlinear, and underactuated, given by

$$\dot{q} = g_1(q)v_1 + g_2(q)v_2 + \cdots + g_m(q)v_m \tag{1.1}$$

Here $q \in M$ is the state vector of the system, M is the state space, and $M \subset R^n$, where n is the dimension of the configuration space M, to which the vector q belongs. Vector $v \in R^m$ is the input or the control vector of dimension m. Vectors $g_i(q) \in R^n; i = 1,2,....m$ are vector fields on M and are assumed to be smooth and linear time invariant. The system is called drift-free because the system state does not change under zero-input conditions. Also, the system is underactuated because the dimension of the space spanned by the control vector is less than the dimension of the configuration space.

A special case of Equation 1.1 with two inputs was presented in [1]. In [1] the motion planning tasks for a car-like robot were defined and the feedback control design was studied. The control was achieved using various control strategies for each task. The kinematic-based control of an underwater vehicle is an extension of the work presented in [1]. The extended problem is higher dimensional with four inputs. In this book the controllability of the system is discussed and proved as related to motion planning. We present feedback control laws that give global stabilization of the vehicle about a desired trajectory and about a point. This is achieved by transforming the kinematic into canonical chained and power forms. The book presents the method of converting the kinematic model into chained and power forms via state feedback and coordinate transformation.

For trajectory tracking of underwater vehicles [23] proposed a stable tracking control method based on a Lyapunov function. In [22] and [23] a Lyapunov-like function is used to develop a nonlinear feedback control scheme. The control achieves global stabilization about a desired trajectory. However, the system is not point stabilizable with the use of the proposed controller. In this book we will be making use of the full state feedback (approximate linearization) scheme for trajectory tracking. This scheme results in local asymptotic stabilization only. Exact nonlinear control (full state linearization) design is used to achieve the global stabilization. In this case static state feedback fails to achieve the goal. However, the dynamic state feedback is used to serve the purpose. Here the control design is done using the chained form system.

The kinematic model of an underwater vehicle belongs to a class of systems that cannot be stabilized by a pure state feedback law [2]. The design of globally asymptotic stabilizing controllers for nonholonomic systems is challenging. The design is difficult in a sense that no time-invariant smooth static stabilizing controller exists for such systems [2]. Various control schemes have been adopted for this purpose.

One way to deal with this is to use time-varying, smooth controllers. This approach has been extensively studied in [3] and [4]. In [3] it is shown that time-varying smooth control laws for driftless systems have necessarily algebraic (not exponential) convergence rates. Asymptotic stabilization for underwater vehicles using time-varying smooth feedback laws was achieved in [4].

Another alternative is the use of the nonsmooth feedback controllers that can achieve exponential convergence. These schemes have been proposed in [5] and [6]. In [25] a discontinuous piecewise smooth control law was proposed and exponential convergence to a constant desired configuration was achieved. In [15] a nonsmooth time-invariant controller was proposed to achieve the exponential convergence with stability to a constant desired configuration. The controller was implemented using the chained form. In this book, we have adopted the former approach, that is, we used a time-varying and smooth feedback. The control design method for stabilization used herein is adopted from [7]. To this end, a transformation of the kinematic model into power form is derived. The controller achieves global stabilization to a constant configuration for an underwater vehicle.

The latter part of the book addresses the feedback control and robust control design for a dynamic model of an underwater vehicle. The motion of underwater vehicles is presented as a kinematic and a dynamic model. The three-dimensional vehicle motion may be described in terms of the twelve nonlinear system equations:

$$
\begin{aligned}
M\frac{dv(t)}{dt} &= f(v,q)+g(v,q)u(t) \\
\frac{dq(t)}{dt} &= h(v,q)
\end{aligned}
\tag{1.2}
$$

The first equation is the dynamic model of the system, and the second is the kinematic model. The vector $v(t)$ is the velocity vector and $q(t)$ is the position vector in the fixed frame coordinates. In order to design the feedback control for point stabilization of the dynamic model, the control methodology is nonlinear feedback linearization. The design technique adopted for the dynamic control is *backstepping*. The book also discusses the feedback control for these models in the presence of uncertainties where the goal is to design the controllers to minimize the effect of uncertainties. The control is achieved using Lyapunov redesign and *robust backstepping* for uncertain models. In both methods we achieve the control design for the kinematic model first, which is then used to achieve the overall control design of the dynamic model. The outline of the chapters is summarized below.

Chapter 2 gives an introduction and general overview of the motion planning of autonomous vehicles. The concepts of nonholonomy, underactuated systems, and a kinematic model of the nonholonomic systems, and some examples are shown. Then the general problem of motion planning and the related issues are formulated for a class of the nonholonomic systems, with a review of some particular applications.

Chapter 3 presents an overview and detailed analysis of the related motion planning tasks of an autonomous underwater vehicle. The chapter presents in detail the derivation of the mathematical modeling of the system. For motion planning tasks, the kinematic model of the system is obtained and the issues related to nonlinear controllability of the system are studied in detail. Finally, for the purpose of control design, the system is converted into a chained form. The method of converting a multi-input nonholonomic system into a chained form is also discussed.

Chapter 4 presents the control design and the simulation results obtained for the model of an underwater vehicle developed in Chapter 3. The feedback control design is developed using the kinematic model of the system. The performance of the controllers using various techniques of control design is obtained and evaluated for different motion planning tasks, such as trajectory tracking, point stabilization, and path following. The chapter also presents the simulation results obtained for different controllers. The simulation results are used to compare and evaluate the performance of the various controllers for different paths following tasks.

Chapter 5 presents the control design of the dynamic model of an underwater vehicle. The control design is obtained and evaluated for a motion planning task of point stabilization. The control design technique adopted here is feedback linearization using backstepping for a dynamic model. The problem of control and stability is formulated and sufficient conditions for Lyapunov stability for the designed controllers are derived.

Chapter 6 presents a design of robust nonlinear feedback controllers for the dynamic model representing underwater vehicles. We discuss the feedback control for these models in the presence of uncertainties where the goal is to design the controllers to minimize the effect of uncertainties on the performance of the system. The robust controllers are designed for point stabilization using Lyapunov redesign and robust backstepping methods.

1.2 EXAMPLES OF UNDERWATER VEHICLE CONSTRUCTION

This section provides a brief overview of the construction of underwater vehicles. Readers are encouraged to study [33] and [34] for more details.

Underwater vehicles have two main issues to deal with. They need to be able to stay underwater and to maneuver. The vehicles can be designed so that they stay underwater, so that their body weight is water neutral, or they can use actuation to actively stay underwater. They also need actuation to be able to go underwater from the water surface, move around underwater, and also come back up to the surface.

Submarines can be dynamic or static. Dynamic submarines usually will float on water and are actively pushed down by propellers to keep them underwater. Static submarines, on the other hand, add water into a submarine chamber (called a ballast tank) to add weight to it, so the submarine can get heavier. When it wants to come up, it uses compressed gas to push the water out, so that the submarine becomes light again.

Remote-controlled (RC) submarines, which can be purchased in many stores, have a motor that drives a propeller at the back that pushes the vehicle forward, and rudder and stern (and sail) planes to enable the submarines to produce yaw motion and pitch motion. Rudders enable the yaw motion by their rotation, and stern planes produce the pitch motion just like elevators do in RC planes. These actuators are shown in Figure 1.1. A more detailed view of the rudder and stern is shown in Figure 1.2.

The weight distribution keeps the submarine facing right side up by keeping the heavier weight at the bottom, not the top. The sail plane with the stern plane keeps the submarine balanced. The electronics is kept inside the water-tight container (WTC) and contains two servos and a DC motor. One servo controls the rudder angle and the other controls the stern. The location of the WTC and the weight distribution are shown in Figure 1.3.

The WTC generally has three compartments. The first compartment will have the battery and speed controller, the second will have a ballast system, and the third the servos. Typically, the ballast system will have a tank with pressurized air and valves. The valves are used to allow water to flood in, to increase the weight of the submarine, therefore enabling it to go down, or to let water come out when the pressurized air is let into the chamber, so that the weight of the submarine is reduced, to enable it to rise. The

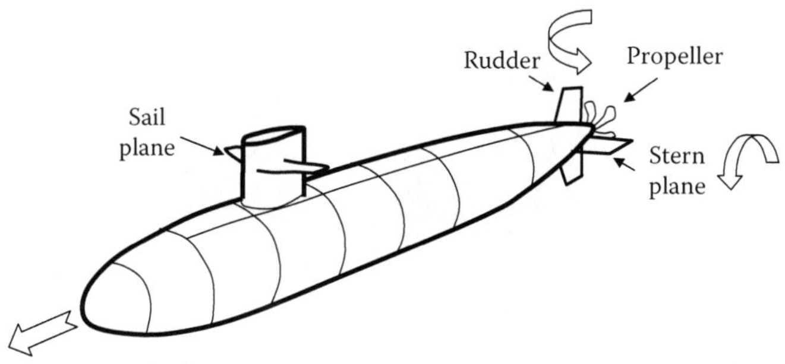

FIGURE 1.1 Submarine actuators.

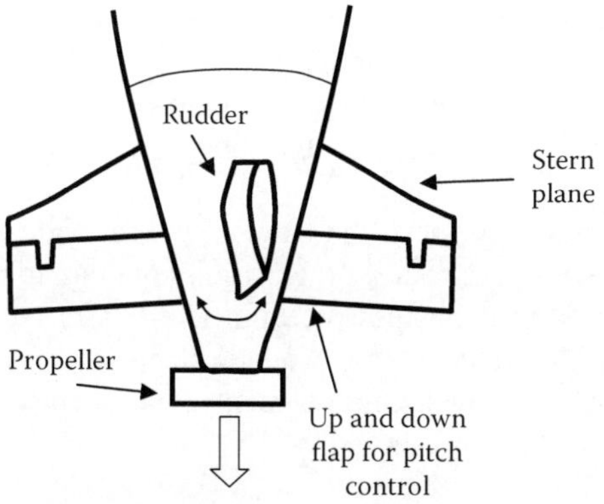

FIGURE 1.2 Submarine rudder and stern.

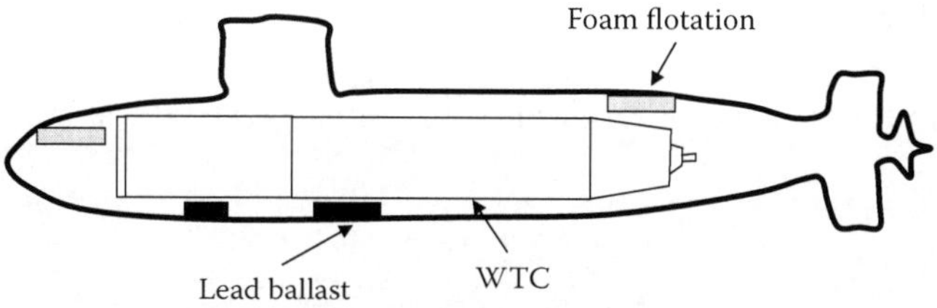

FIGURE 1.3 WTC location and weight distribution.

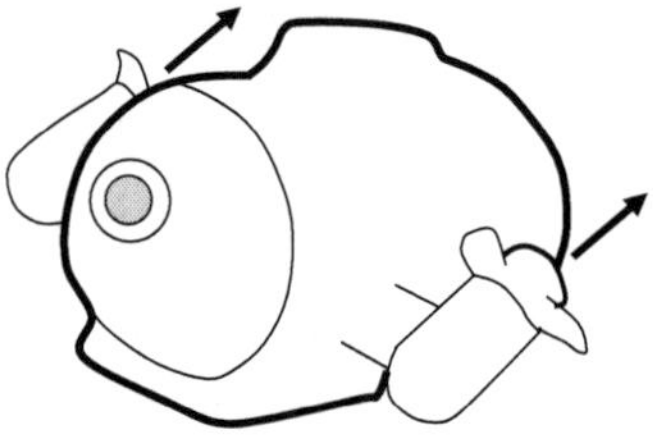

FIGURE 1.4 Twin-propeller-based dynamic RC submarine.

other electronic components that are usually present are the speed controller connected to the radio signal receiver, which receives the signal from the remote control to control the motor and servos; the pitch control using an accelerometer to control the pitch angle; and a microsafe, a device that turns the ballast system off if the radio signal is lost for some time, so that the submarine will float on top in case of a lost radio signal.

Submarines communicate underwater with sonar signals using audio frequencies. Submarines also use GPS for navigation, but that signal is lost underwater; therefore, an underwater inertial navigation system (INS) is used that is composed of accelerometers and gyroscopes to measure how the submarine is moving so that it can keep track of its position and orientation.

Dynamic submarines don't have the ballast system, and therefore they have to be actively forced down in the water; otherwise, they float up on the surface. Many inexpensive dynamic RC submarines are built using twin motors that are used to dive, surface, turn, etc. The angle of the propellers can be changed before putting the submarine into water to get different movements. One such submarine is illustrated in Figure 1.4.

1.2.1 Propeller Principle

Since underwater vehicles have propellers, this section will provide a brief description of the theory of propeller action. Propeller thrust is produced due to the forces that are created by the "wings" that are part of the propellers. We present the fundamentals of how force is produced in the wings, and then how propellers produce thrust because of them.

1.2.1.1 Wings

To understand how the air flowing around a wing section produces lift, let us study Figure 1.5. The figure shows fluid flowing around a wing section. The streamline (a line

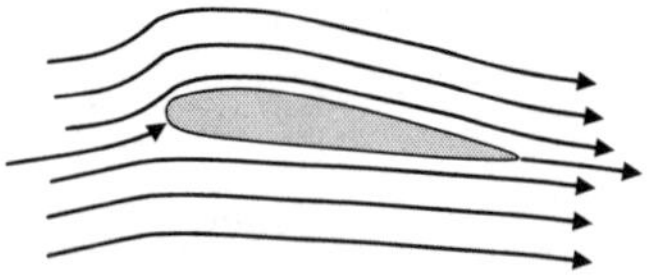

FIGURE 1.5 Airflow around a wing section (Kutta condition).

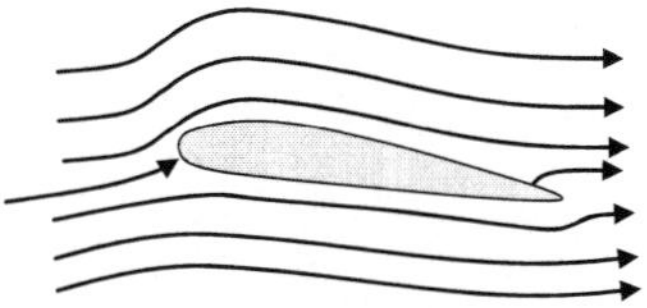

FIGURE 1.6 Noncirculatory airflow around a wing section.

whose tangent shows the direction of flow at every point) on the left that ends with an arrow on the wing surface is the one that reaches the stagnation point on the wing. This means that above that the fluid flows on top of the wing, and below that point the fluid flows under the wing. The figure shows that the fluid from the top and bottom will have the same magnitude and direction at the trailing edge. This is a condition (Kutta condition [33]) that creates lift from fluid flowing around a wing at some angle of attack. If we place a wing in a fluid flow, we can expect the streamlines to be as shown in Figure 1.6. This fluid flow, however, has no circulation. Without fluid circulation, there can be no lift. In fact, the lift on a wing is proportional to the air circulation about the wing (Kutta–Joukowski theorem [34]).

According to the Kutta–Joukowski theorem, the lift per unit length of a wing is given by

$$Lift = airspeed \times air\ circulation \times air\ density$$

The fluid flow that follows the Kutta condition (the trailing edge boundary condition) can be obtained by superimposing a circulatory airflow on the noncirculatory fluid flow. This is shown in Figure 1.7.

There is another theorem (Kelvin's circulation theorem [33]) that says that the rate of change of circulation around a closed curve that consists of the same fluid material is zero. When a vehicle such as an airplane starts accelerating, it does not follow the Kutta condition, and consequently, it does not have circulation. After the speed increases further, the Kutta condition is satisfied (creating circulation around the wings), and to make the total circulation zero, another circulating air is created behind the wing (called the *wake*). This is shown in Figure 1.8.

Due to the addition of clockwise circulation that is flowing in the direction of the regular airflow, the speed on top of the wing is higher than the fluid speed below the wing. Bernoulli's theorem shows that where the airspeed is high, there the pressure is low, and where the airspeed is low, there the air pressure is high [34]. Since the air

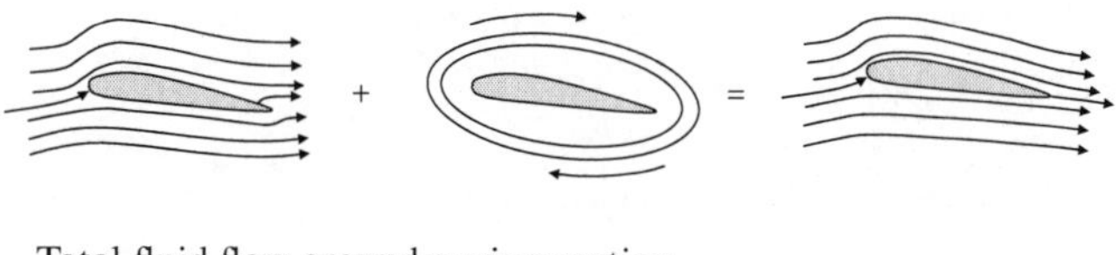

FIGURE 1.7 Total fluid flow around a wing section.

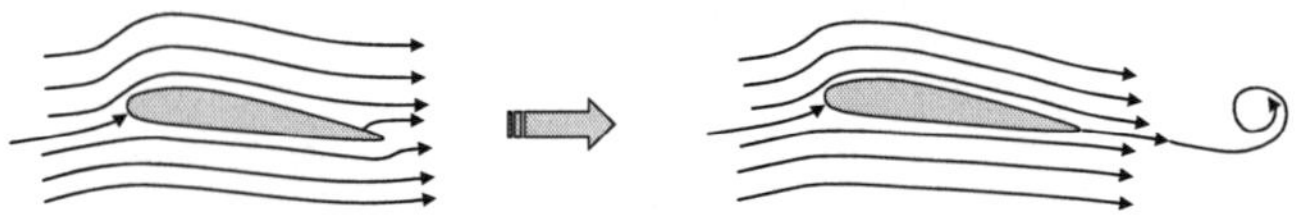

FIGURE 1.8 Total fluid circulation around a wing section.

pressure is higher at the bottom of the wing than at the top, there is an upward force (lift) produced on the wing. This is shown in Figure 1.9.

Now when we look at a wing with a finite span, such as on an underwater vehicle for stabilization, because of low pressure on top of the wing, the fluid from the bottom starts circulating to the top, and the whole circulation travels behind the plane. This is shown in Figure 1.10.

The lift force that is generated by the wings is related to the angle of attack (i.e., the angle that the fluid velocity at a distance in front of the wing makes with the wing chord line). When the angle of attack is at zero degrees, there is no lift produced. As the angle of attack is increased, the lift produced increases, up to some angle of attack, after which further increase in the angle decreases the lift. This is shown in Figure 1.11.

1.2.1.2 Propellers

Propellers are extremely important for robotic underwater vehicles. They provide the thrust for forward motion as well as for lift. Figure 1.12 shows a three-blade propeller. Propellers can come in many blade configurations, such as two blade, three blade, and four blade.

Propellers produce forward thrust by accelerating fluid back. The amount of fluid forced back depends on the speed of rotation of the propeller as well as at its pitch angle. With a small pitch angle, the amount of fluid pushed back is small, and therefore less forward thrust is produced. However, in that case the propeller can rotate faster. With a high pitch angle, the blades rotate slower but produce more thrust by pushing more fluid back. Propellers can be designed to have fixed pitch, or can be made such that the pitch of the propeller can be changed in real time (by an actuator). The concept of variable pitch can be compared to having gears in a motor. The pitch could be controlled automatically (using some electronics) to produce a constant speed of rotation.

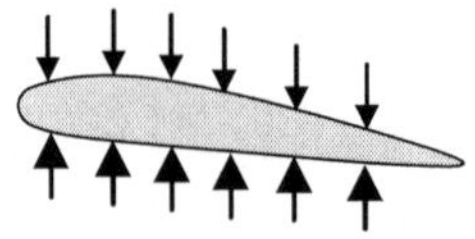

FIGURE 1.9 Lift due to pressure difference produced on a wing section.

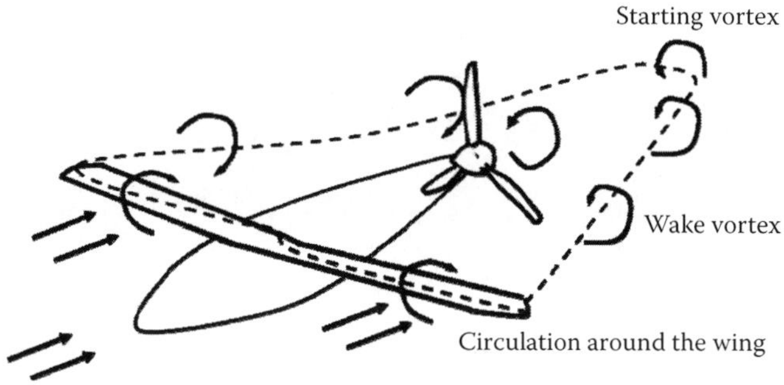

FIGURE 1.10 Vortex.

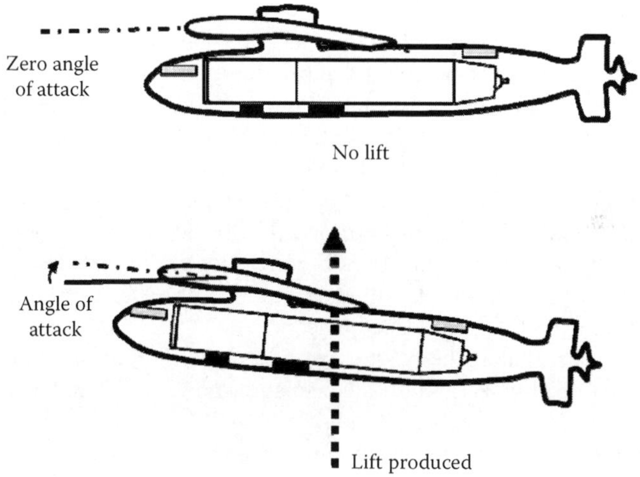

FIGURE 1.11 Angle of attack and lift.

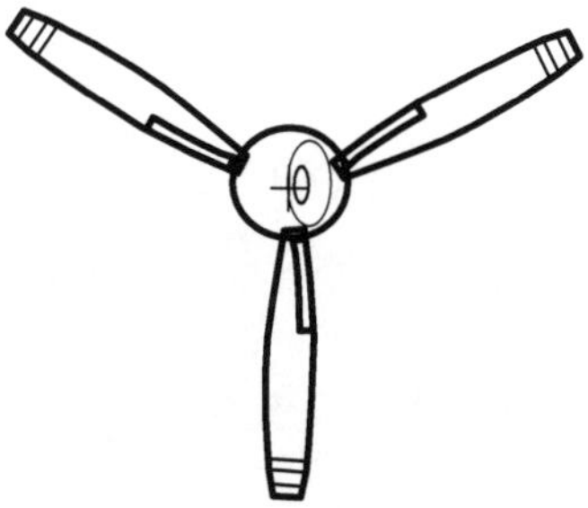

FIGURE 1.12 Propeller.

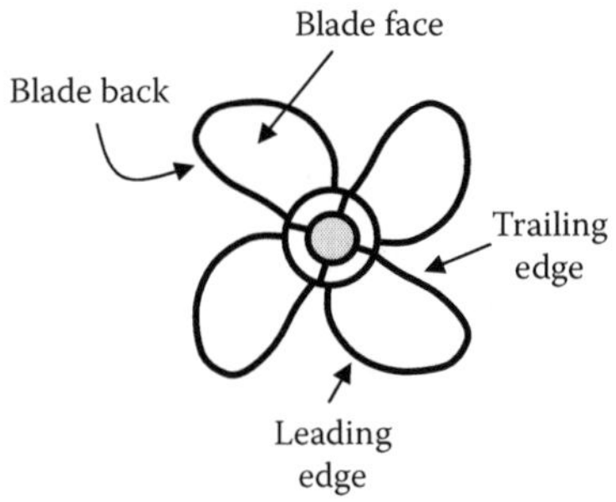

FIGURE 1.13 Propeller parts.

Another way to get a good estimate of the propeller thrust is to apply the theory of the wing section to the propeller blade to figure out the lift force that is generated on the blade as the fluid flows across it. The fluid speed that any part of the propeller blade is subjected to depends on the distance of that part from the center of rotation. Therefore, the propeller blades are twisted since at different radii, the relative speeds are different, and that makes the angle of attack different, if the blade is not designed to be twisted.

Let us look at the parts of a propeller in more detail (see Figure 1.13). The leading edge of the propeller is where the fluid first cuts through the blade. The trailing edge is where the fluid exits the propeller surface. The blade face is the side of the blade that pushes the fluid away, creating a positive pressure on the surface. The blade back is the back side of the blade, which gets a negative pressure as the fluid is moved away from that region by the rotating blade.

As the propeller rotates, the blade face pushes the fluid back and the back side starts pulling the fluid in (push–pull mechanism), similar to a fan, and this creates a forward thrust from the propeller (see Figure 1.14).

As the propeller makes one revolution, the geometric pitch is the distance the propeller would have moved through a soft material like softwood if it were a screw. The effective pitch (smaller than the geometric pitch) is the actual distance

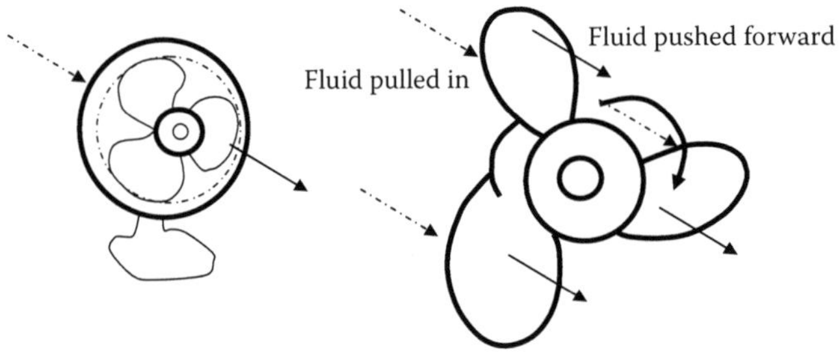

FIGURE 1.14 Propeller thrust.

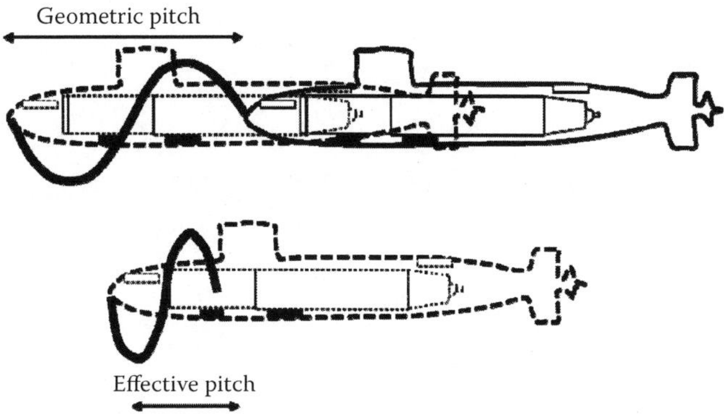

FIGURE 1.15 Propeller pitch.

the plane or a boat (or any other robot) moves. This difference is due to the propeller slippage through the fluid (see Figure 1.15).

1.2.2 Commercially Available Underwater Vehicles

Autonomous underwater vehicles (AUVs) have been designed in laboratories and research centers since around the 1950s, but there are only a few companies that sell these commercially. Chief among these are Kongsberg Maritime (http://www.km.kongsberg.com) and Bluefin Robotics Corporation (http://www.bluefinrobotics.com). Below is a brief description of the AUV models available from these manufacturers.

Kongsberg manufactures the HUGIN AUVs and REMUS AUVs, besides providing various instruments and software for navigation. The HUGIN AUV can be used for subsea sensing. These AUVs are designed to go as deep as 4,500 meters (HUGIN 4500), and can propel and navigate themselves. They also have the ability to handle disturbances. The REMUS AUVs have the advantage that they can be custom designed. These AUVs at present are available in three models: REMUS 100, REMUS 600, and REMUS 6000. REMUS 100 is lightweight and can go 100 meters deep. REMUS 600 has the ability to go to 3,000 meters deep, while REMUS 6000 has a maximum operation depth of 6,000 meters.

Bluefin Robotics has been associated with MIT's Autonomous Underwater Vehicle Laboratory. The different models available are Bluefin-9, Bluefin-12, Bluefin-21, and Bluefin-21 BPAUV. These AUVs are designed for low as well as deep water applications. The numbers in each of the model names refer to the AUV diameter in inches. These AUVs can carry varying payloads and therefore can vary in length.

Other manufacturers of AUVs include International Submarine Engineering Ltd. (http://www.ise.bc.ca/) and GAVIA Autonomous Underwater Vehicles (http://www.gavia.is).

Readers can check the latest models and specifications from the individual manufacturers' Web sites.

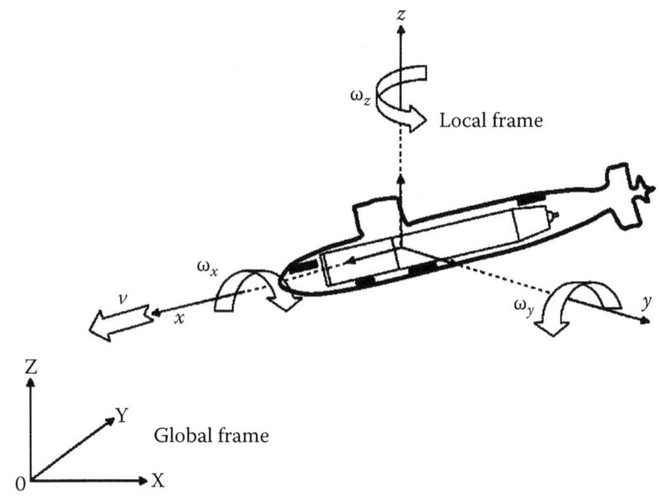

FIGURE 1.16 World frame.

1.3 VEHICLE KINEMATICS FUNDAMENTALS

There are essentially six degrees of freedom that the vehicle has: three that indicate the position, and the other three that indicate the orientation. These six degrees of freedom identify the state of the vehicle. We consider the state of the vehicle with respect to some general frame of reference for the world (Figure 1.16).

Although there are six degrees of freedom for the vehicle to be placed with respect to position and orientation, there are typically only four variables that can be directly controlled. These four variables can be considered more naturally in the body moving frame (Cartan) of reference. In the frame of reference that is attached to the body of the vehicle, as shown in Figure 1.16, the vehicle can move forward in the direction of the body x axis, and can also produce rotation about the body x, y, and z axes. We consider the control variables to be the velocities in these four body variables, that is, the linear velocity in the direction of the body x axis and the three angular velocities. We need a way to relate these control variables to keep track of the state of the vehicle. We present the Frenet–Serret equations to develop this relationship. Note that, in practice, the actual actuation is obtained through controlling the actuators that produce those control velocities through the dynamics of the system. We assume that there are local feedback loops already developed that can control the linear and angular body velocities, and the aim of the controller is to control the motion of the vehicle in the global frame.

1.3.1 Frenet–Serret Equations for Cartan Moving Frame

Let's consider a point fixed on the vehicle. We can assume the point to be the center of gravity for the vehicle body, or any convenient fixed point on the vehicle body. The position of this point is a function of time that maps time to a point in R^2.

Definition: A regular parametric representation of the position vector in terms of the parameter s is a function $x(s) \in C^1(-\alpha,\alpha)$ such that $\frac{dx}{ds} \neq 0, \forall s \in (-\alpha,\alpha)$.

Definition: A real valued function $t(s) \in C^1(-\alpha,\alpha)$ such that $\frac{dt}{ds} \neq 0, \forall s \in (-\alpha,\alpha)$ is called an allowable change of parameter.

Regular parametric representations divide curves into equivalent classes, since an allowable change of a parameter's relationship is reflexive, symmetric, and transitive, that is, it is an equivalence relationship.

Definition: A regular curve is an equivalence class of regular parametric representations.

Given a curve parameterized by arc length

$$x(s) : (-\alpha,\alpha) \rightarrow R^2$$

that represents the motion of the vehicle, its unit tangent vector is given by

$$t(s) = \dot{x}(s)$$

The plane normal to the curve at point x is called the *normal plane*, as shown in Figure 1.17.

It is easy to verify that $\|t(s)\| = 1$. This implies that the inner product of $t(s)$ with itself is 1, that is,

$$<t(s), t(s)> = 1$$

Differentiating this equation with respect to the arc length variable s, we get

$$<\dot{t}(s), t(s)> + <\dot{t}(s), t(s)> = 0$$

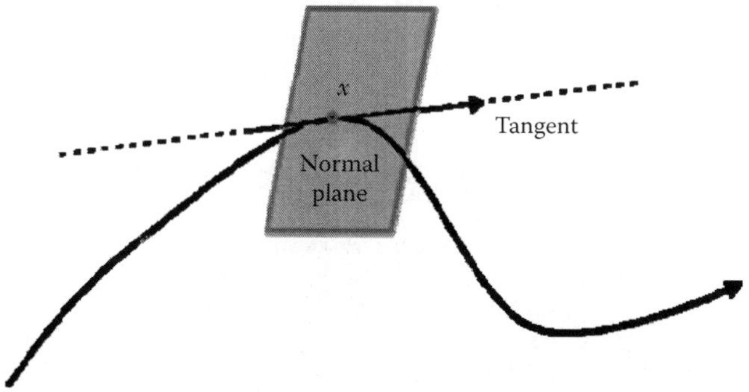

FIGURE 1.17 Normal plane.

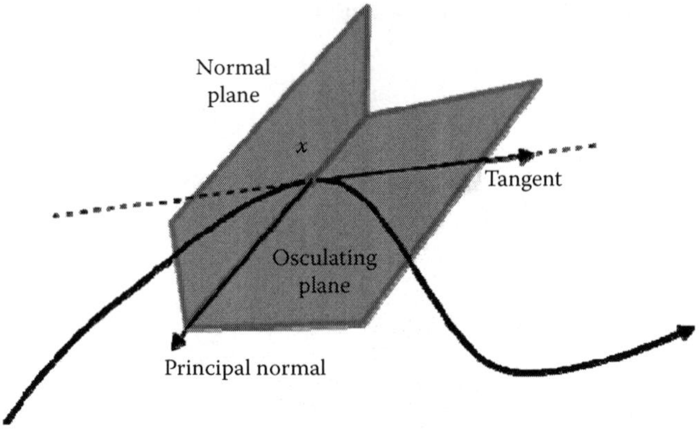

FIGURE 1.18 Normal and osculating planes.

Since the inner product in real vector space is commutative, we get

$$<\dot{t}(s), t(s)> = 0$$

which implies that $\dot{t}(s) \perp t(s)$. The curvature of the curve, $\kappa(s)$, is the norm of the tangent vector. Let the unit vector in the direction of $\dot{t}(s)$ be called $n(s)$. Hence, we get

$$\dot{t}(s) = \kappa(s) n(s)$$

The plane containing the vectors $t(s)$ and $n(s)$ is called the *osculating plane*, as shown in Figure 1.18.

We have that $\|n(s)\| = 1$. This implies that the inner product of $n(s)$ with itself is 1, that is,

$$<n(s), n(s)> = 1$$

Differentiating this equation with respect to the arc length variable s, we get

$$<\dot{n}(s), n(s)> + <n(s), \dot{n}(s)> = 0$$

Since the inner product in real vector space is commutative, we get

$$<\dot{n}(s), n(s)> = 0$$

which implies that $\dot{n}(s) \perp n(s)$.

We define the binormal to the curve as

$$b(s) = t(s) \times n(s)$$

which also gives us

$$n(s) = b(s) \times t(s)$$

Differentiating the equation gives us

$$b(s) = \dot{t}(s) \times n(s) + t(s) \times \dot{n}(s)$$

This gives, after substitution,

$$\dot{b}(s) = \kappa(s)\,(n(s) \times n(s)) + t(s) \times \dot{n}(s)$$

Since the cross product of a vector with itself is zero, this gives

$$\dot{b}(s) = t(s) \times \dot{n}(s)$$

The plane containing the vectors $t(s)$ and $b(s)$ is called the *rectifying plane*, as shown in Figure 1.19.

Since $\dot{n}$ is perpendicular to n, we have

$$\dot{n}(s) = \mu(s)t(s) + \tau(s)b(s)$$

This gives us

$$\dot{b}(s) = t(s) \times [\mu(s)t(s) + \tau(s)b(s)]$$

Again, since a cross product of a vector with itself is zero, we get

$$\dot{b}(s) = \tau(s)[t(s) \times b(s)] = -\tau(s)n(s)$$

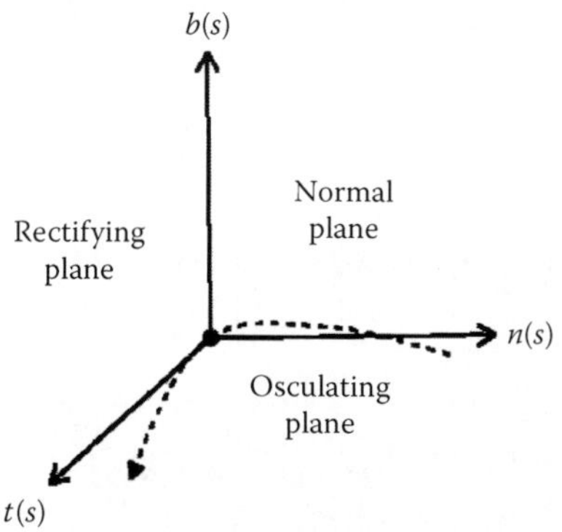

FIGURE 1.19 Normal, osculating, and rectifying planes.

Since $n(s) = b(s) \times t(s)$, we get by differentiating

$$\dot{n}(s) = \dot{b}(s) \times t(s) + b(s) \times \dot{t}(s)$$

which after substitutions gives

$$\begin{aligned} \dot{n}(s) &= -\tau(s)[n(s) \times t(s)] + \kappa(s)[b(s) \times n(s)] \\ &= \kappa(s)t(s) + \tau(s)b(s) \end{aligned}$$

Hence, in summary we have

$$\begin{aligned} \dot{x}(s) &= t(s) \\ \dot{t}(s) &= \kappa(s)n(s) \\ \dot{n}(s) &= \kappa(s)t(s) + \tau(s)b(s) \\ \dot{b}(s) &= -\tau(s)n(s) \end{aligned}$$

The vectors $t(s)$, $n(s)$, and $b(s)$ are all orthogonal to each other; therefore, the matrix containing them as the column vectors belongs to a matrix group $SO(3)$.

$$[t(s), n(s), b(s)] \in SO(3).$$

Thus,

$$\begin{bmatrix} t(s) & n(s) & b(s) & x(s) \\ 0 & 0 & 0 & 1 \end{bmatrix} \in SE(3)$$

which gives us

$$\frac{d}{dt}\begin{bmatrix} t(s) & n(s) & b(s) & x(s) \\ 0 & 0 & 0 & 1 \end{bmatrix} = \begin{bmatrix} t(s) & n(s) & b(s) & x(s) \\ 0 & 0 & 0 & 1 \end{bmatrix}$$

$$\begin{bmatrix} 0 & -\kappa(s) & 0 & 1 \\ \kappa(s) & 0 & \tau(s) & 0 \\ 0 & -\tau(s) & 0 & 0 \\ 0 & 0 & 0 & 0 \end{bmatrix}$$

The above equations are called the *Frenet–Serret equations of a curve*. The evolution of this curve in R^2 is given by

$$\dot{g} = gx$$

where $g \in SE(3)$ group and X is an element of the Lie algebra $se(3)$. We can regard the curvature $\kappa(s)$ and the torsion $\tau(s)$ as inputs to the system, so that if

$$u_1(s) = \kappa(s)$$
$$u_2(s) = -\tau(s)$$

then

$$\dot{g} = g\begin{bmatrix} 0 & -u_1 & 0 & 1 \\ u_1 & 0 & -u_2 & 0 \\ 0 & u_2 & 0 & 0 \\ 0 & 0 & 0 & 0 \end{bmatrix}$$

which is the special case of the general form describing the state evolution of a left invariant control system in $SE(3)$.

An example of the general form of a left invariant control system in $SE(3)$ is given by an aircraft flying in R^2.

$$\dot{g} = g\begin{bmatrix} 0 & -u_3 & u_2 & u_4 \\ u_3 & 0 & -u_1 & 0 \\ -u_2 & u_1 & 0 & 0 \\ 0 & 0 & 0 & 0 \end{bmatrix}$$

The inputs u_1, u_2, and u_3 control the roll, pitch, and yaw of the aircraft, and the input u_4 controls the velocity in the forward direction.

The special case of above in $SE(2)$ is an example of a unicycle rolling on the plane.

$$\dot{g} = g\begin{bmatrix} 0 & -u_2 & 1 \\ u_2 & 0 & 0 \\ 0 & 0 & 0 \end{bmatrix}$$

In this case, the input u_2 controls the angle of the wheel.

The previous formulation describes kinematic steering problems, as it is assumed that we have direct control of the velocities of the rigid bodies. However, in physical systems, we only have control of the forces and torques that drive the motion of the body. The more realistic approach then is to formulate the steering problem using a dynamic model of the system, where the forces and torques are the inputs. The control of the dynamic models is much harder to achieve than the kinematic models. In the rest of this book the focus will be to address the control design of both the models for an autonomous underwater vehicle.

1.3.2 Mathematical Background for Rigid Motion in a Plane

We notice that for an underwater vehicle that has a propeller to thrust it forward and has three more actuators to control the three rotations, roll, pitch, and yaw, we can use Frenet–Serret equations to represent the kinematic equations for the motion. We will study these kinematics and rigid motion of the vehicle in more detail in this section. First, we start with rotations.

We will first study three interpretations of a rotation matrix that we will derive in this section, followed by the general rigid motion in a plane. Rotations and translations constitute rigid motions since they do not change the length of vectors that they transform. However, we do not want to allow reflections as rigid motions.

1.3.2.1 Rotation of a Vector

We will derive a rotation matrix that is an operator that works on vectors and rotates them without changing the length.

Figure 1.20 shows a rotation of angle θ of the x axis counterclockwise. Hence, the vector $(1,0)'$ has been transformed into $(\cos\theta, \sin\theta)'$. This can be written as

$$\begin{bmatrix} \cos\theta \\ \sin\theta \end{bmatrix} = R_\theta \begin{bmatrix} 1 \\ 0 \end{bmatrix}$$

Similarly, for the y axis, when rotated by the same angle, we get

$$\begin{bmatrix} -\sin\theta \\ \cos\theta \end{bmatrix} = R_\theta \begin{bmatrix} 0 \\ 1 \end{bmatrix}$$

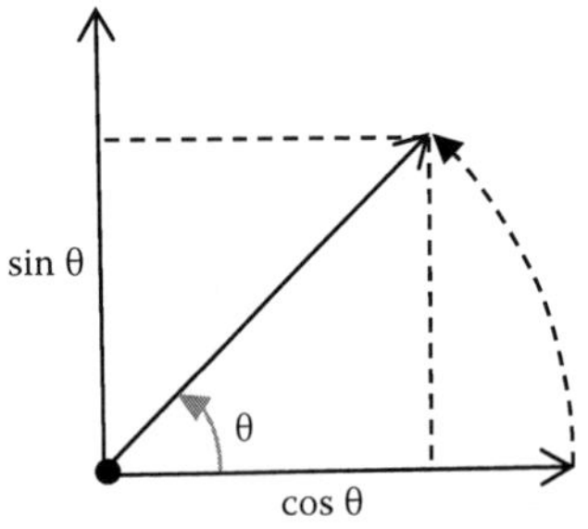

FIGURE 1.20 Rotation in a plane of the x axis.

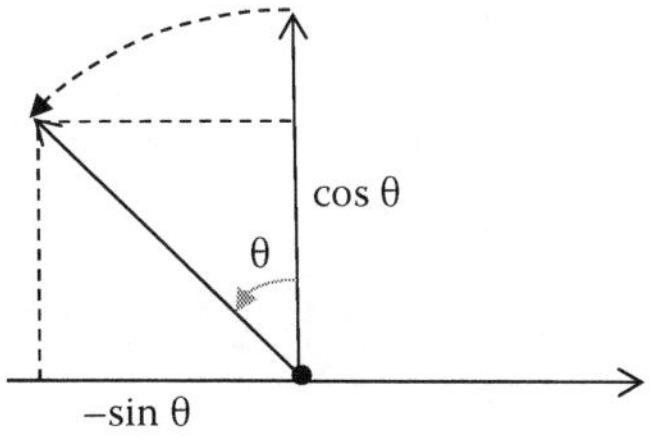

FIGURE 1.21 Rotation in a plane of the y axis.

This is shown in Figure 1.21.

Combining these two results, we get the matrix representation for the linear rotation operator that rotates the vector counterclockwise by angle θ:

$$R_\theta = \begin{bmatrix} \cos\theta & -\sin\theta \\ \sin\theta & \cos\theta \end{bmatrix}$$

1.3.2.2 Vector Represented in a Rotated Frame

This interpretation of the rotation matrix is explained using Figure 1.22. Given a vector V, its representation in R^2 is transformed to its representation in the frame F_2, by using the rotation matrix as follows:

$$v^{F_1} = R_\theta v^{F_2}$$

Here, the vector v^{F_1} is the representation of the vector in R^2 in frame F_1; the vector v^{F_2} is the representation of the vector in R^2 in frame F_2. It is clear that the angle between the vector and the x axis of the frame gets rotated by angle θ.

1.3.2.3 Representation of a Rotated Frame

In this interpretation the rotation matrix represents the axis of one frame of reference with respect to another. Refer to Figure 1.23.

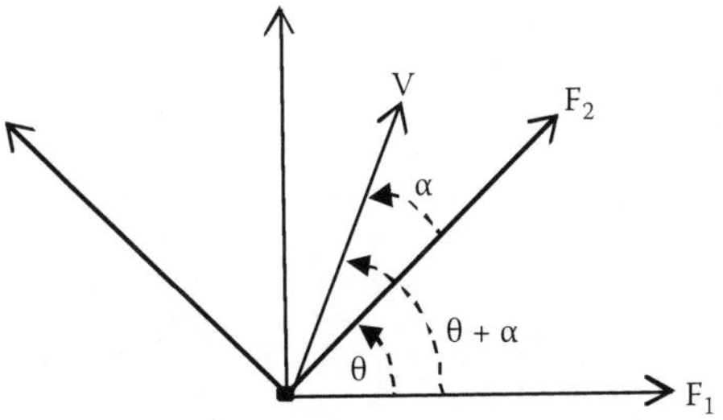

FIGURE 1.22 Rotation of the frame for the vector.

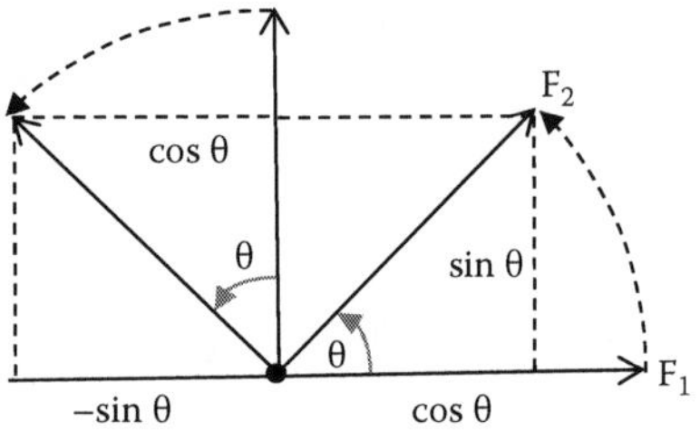

FIGURE 1.23 Rotation of the frame.

The vector representation of the x axis of the frame F_2 is

$$x_{F_2} = \begin{bmatrix} \cos\theta \\ \sin\theta \end{bmatrix}$$

And the vector representation of the y axis of the frame F_2 is

$$y_{F_2} = \begin{bmatrix} -\sin\theta \\ \cos\theta \end{bmatrix}$$

Combining the two we get the rotation matrix. If the unit vectors in the frame F_1 are given by vectors i_1 and j_1, and those of the frame F_2 are given by i_2 and j_2, then the rotation matrix, as shown here, is

$$R_\theta = \begin{bmatrix} i_2.i_1 & j_2.i_1 \\ i_2.j_1 & j_2.j_1 \end{bmatrix}$$

1.3.2.4 Group Representation

Since rotations are additive, we can represent them as a group. For the definition of a group refer to the next section. We can use an exponentiation notation for the rotation, as given by

$$e^{\vec{z}\theta} = R_\theta = \begin{bmatrix} \cos\theta & -\sin\theta \\ \sin\theta & \cos\theta \end{bmatrix}$$

Here, the vector $\vec{z}$ in the power indicates that the rotation is about that axis. Another nice motivating example for using exponential representation is as follows. First, we review the vector cross product. In order to do that, we start using vectors in R^3. Given two vectors, their cross product is shown below.

$$a = \begin{bmatrix} a_1 \\ a_2 \\ a_3 \end{bmatrix} = a_1 i + a_2 j + a_3 k\,, \quad b = \begin{bmatrix} b_1 \\ b_2 \\ b_3 \end{bmatrix} = b_1 i + b_2 j + b_3 k$$

where

$$i = \begin{bmatrix} 1 \\ 0 \\ 0 \end{bmatrix}, \; j = \begin{bmatrix} 0 \\ 1 \\ 0 \end{bmatrix}, \; k = \begin{bmatrix} 0 \\ 0 \\ 1 \end{bmatrix}$$

then

$$a \times b = (a_2 b_3 - a_3 b_2) i + (a_3 b_1 - a_1 b_3) j + (a_1 b_2 - a_2 b_1) k$$

This operation can be represented conveniently in a determinant notation as follows:

$$a \times b = \begin{vmatrix} i & j & k \\ a_1 & a_2 & a_3 \\ b_1 & b_2 & b_3 \end{vmatrix}$$

However, more importantly, the cross product by a fixed vector can be given a matrix operator form as follows. According to this, a cross product by a vector is equivalent to a matrix operation.

$$a \times b = [a \times] b = \begin{bmatrix} 0 & -a_3 & a_2 \\ a_3 & 0 & -a_1 \\ -a_2 & a_1 & 0 \end{bmatrix} \begin{bmatrix} b_1 \\ b_2 \\ b_3 \end{bmatrix}$$

The matrix $[a \times]$ is skew symmetric since $[a \times]^T + [a \times] = 0$, where T denotes transpose and zero is the matrix with all zero-valued terms.

Consider the position vector of a particle rotating in a circle with angular velocity ω, as shown in Figure 1.24; then the velocity vector is given by

$$\dot{p}(t) = \omega \times p(t) = [\omega \times] p(t)$$

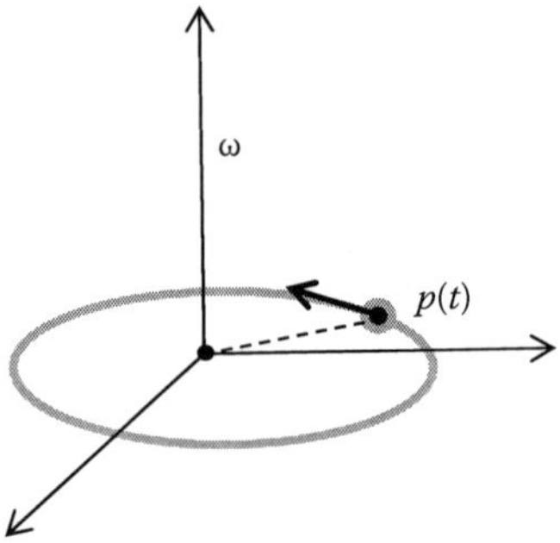

FIGURE 1.24 Rotation of the position vector.

This is a linear differential equation, whose solution is

$$p(t) = e^{[\omega \times]t} p(0)$$

For this problem, we know that the position is given in terms of the rotation matrix. Hence, we have

$$p(t) = e^{[\omega \times]t} p(0) = e^{\vec{\omega}(\omega t)} p(0)$$

An infinitesimal rotation can be obtained by taking a very small angle increment in rotation. So, if we take the rotation angle to be very small, then we get $\sin\theta \cong \theta$ and $\cos\theta \cong 1$, and therefore the rotation is

$$R_\theta = \begin{bmatrix} 1 & -\theta \\ \theta & 1 \end{bmatrix}$$

So, an incremental change from zero rotation gives the tangent space of skew-symmetric matrices by taking θ and finding the incremental rotation difference as

$$[R_{\omega(t+\Delta t)} - R_{\omega t}]/\Delta t = [R_{\omega t} R_{\omega \Delta t} - R_{\omega t}]/\Delta t = R_{\omega t}[R_{\omega \Delta t} - I]/\Delta t = R_{\omega t} \begin{bmatrix} 0 & -\omega \\ \omega & 0 \end{bmatrix}$$

1.3.2.5 Homogeneous Representation

To represent affine transformation, that is, rotation and translation for a two-dimensional motion for a vehicle, we can use homogeneous representation. Observe the translation and rotation of a frame in Figure 1.25.

Any vector that is rotated and then translated is transformed as follows:

$$\begin{bmatrix} v \\ 1 \end{bmatrix} = \begin{bmatrix} R_\theta & \ell_{ab} \\ 0 & 1 \end{bmatrix} \begin{bmatrix} v_0 \\ 1 \end{bmatrix}$$

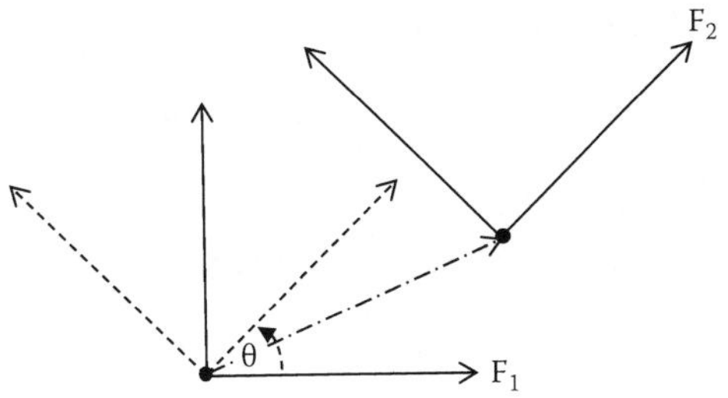

FIGURE 1.25 Rotation and translation.

This shows that the vector v_0 is transformed to v as

$$v = R_\theta v_0 + \ell_{ab}$$

We can develop three-dimensional versions of these transformations, which allow for arbitrary rotations and translations for the vehicle. Details of these are provided in [35].

1.4 LIE GROUPS AND LIE ALGEBRAS

This section covers the mathematical preliminaries that are used for the development, analysis, and control of the kinematic model of an underwater vehicle. The definitions and the mathematical framework presented here will be used throughout the book.

1.4.1 Matrix Groups

Definition (Group): A group G is a set with a binary operation (.): $G \times G \rightarrow G$, such that $\forall$ a, b, c in G, and the following properties are satisfied:

1. Closure: $a \in G, b \in G \Rightarrow a \cdot b \in G$
2. Associativity: $(a.b).c = a.(b.c)$
3. Identity: $\exists$ an identity $e \ni a.e = e.a = a$
4. Inverse: $\exists$ an inverse $a^{-1} \ni a.a^{-1} = a^{-1}.\ a = e$

A group is called *abelian* if $a.b = b.a,\ \forall\ a,b \in G$

Definition (Homomorphism): A homomorphism between groups $\varnothing$: $G \rightarrow H$ is a map that preserves the group operation:

$$\varnothing(a.b) = \varnothing(a).\varnothing(b)$$

Definition (Isomorphism): An isomorphism is a homomorphism that is bijective.

Definition (Field): A field K is a set with two binary operations, addition (+) and multiplication (.), such that:

1. K is an abelian group under (+), with identity 0.
2. $K-\{0\}$ is an abelian group (.), with identity 1.
3. (.) distributes over the (+) $\ni a.(b + c) = a.b + a.c$

Definition (Algebra): An algebra is a vector space with a multiplication that distributes over addition. $M_n(k)$ is an algebra with multiplication defined as the usual multiplication for matrices. For $A, B, C \in M_n(K)$

$$A(B + C) = AB + BC$$

$$(B + C)A = BA + CA$$

Definition (Unit): If A is an algebra, $x \in A$ is a unit if there exists $y \in A$ such that $xy = yx = 1$.

If A is an algebra with an associative multiplication operation and $U \in A$ is the set of units in A, then U is a group with respect to this multiplication operation.

Definition (Matrix group): The class of groups whose elements are $n \times n$ matrices is called a *matrix group.*

Definition (General and special linear groups):

1. The group of units of $M_n(K)$ is the set of matrices M for which $\det(M) \neq 0$, where 0 is the additive identity of K. This group is called the *general linear group* and is denoted by $GL(n,K)$.
2. $SL(n,k) \in GL(n,k)$ is the subgroup of $GL(n,k)$ whose elements have determinant 1. This group, $SL(n,k)$, is called the *special linear group.*

Definition (Orthogonal matrix groups): $O(n,K) \in GL(n,k)$ is the subgroup of $GL(n,k)$ whose element matrices A satisfy the orthogonality condition $\bar{A}^T = A^{-1}$, where $\bar{A}^T$ is the complex conjugate transpose of A. $O(n) = O(n,R)$ is called the *orthogonal group* and $U(n) = U(n,C)$ is called the *unitary group.*

Definition (Special orthogonal matrix groups):

1. $SO(n) = O(n) \cap SL(n,R)$ is the set of all orthogonal matrices of determinant 1. It is called the *special orthogonal group.*
2. $SU(n) = U(n) \cap SL(n,C)$ is the set of all unitary matrices of determinant 1. It is called the *special unitary group.*

Definition (Euclidean matrix groups):

1. The Euclidean group is the set of matrices $E(n)$ such that

$$E(n) = \left\{ A \in R^{(n+1)\times(n+1)} : A = \begin{bmatrix} R & p \\ 0^{1\times n} & 1 \end{bmatrix}, \ R \in GL(n), p \in R^n \right\}$$

2. The special Euclidean group is the set of matrices $SE(n)$ such that

$$SE(n) = \left\{ A \in R^{(n+1)\times(n+1)} : A = \begin{bmatrix} R & p \\ 0^{1\times n} & 1 \end{bmatrix}, \ R \in SO(n), p \in R^n \right\}$$

Definition (Dimension of a matrix group): The dimension of a matrix group G is the dimension of the vector space T of tangent vectors to G at I.

Now let us introduce a family of matrices that will be used to determine the dimensions of matrix groups. Let $so(n)$ denote the set of all skew-symmetric matrices in $M_n(R)$:

$$so(n) = \{A \in M_n(R): A^T + A = 0\}$$

Similarly, the set

$$su(n) = \{A \in M_n(C) : \bar{A}^T + A = 0\}$$

denotes the skew-hermitian matrices, and the set

$$sp(n) = \{A \in M_n(H) : \bar{A}^T + A = 0\}$$

denotes the skew-symplectic matrices. We also define

$$sl(n) = \{A \in M_n(HR): trace(A) = 0\}$$

and

$$se(n) = \left\{ A \in R^{(n+1)\times(n+1)}: \ A = \begin{bmatrix} \hat{w} & p \\ 0 & 0 \end{bmatrix}, \ \hat{w} \in SO(n), p \in R^n \right\}$$

1.4.2 Lie Groups

The discussion of matrix Lie groups needs some definitions from the differential geometry. The material in this section is treated in the context of dynamical control systems in [36] and [37] and from a topological point of view in [38].

Definition (Topological space): A topological space is a set M together with a collection of subsets T of M, that satisfies the following:

1. T contains $\varnothing$ and M.
2. The union of any collection of sets in T is in T.
3. The intersection of a finite collection of sets in T is in T.

The topological space is denoted by (M,T), where T is called a *topology* on M.

Definition (Homeomorphism): Two topological spaces, M and N, are said to be homeomorphic if there exists a function $f: M \rightarrow N$ that is continuous and bijective (i.e., one to one and onto) and whose inverse, f^{-1} is continuous. The function f is called a *homeomorphism.*

Definition (Manifold): An n-manifold is a topological space M such that for any $p \in M$, there exists a neighborhood of p that is homeomorphic to an open set in R^n. The dimension of M is n.

Definition (Smooth manifold): A smooth manifold is a manifold that has an associated maximal atlas.

We now introduce the concept of tangent vectors to manifolds. Let M be a differentiable n-manifold with $p \in M$. Let

$$A(p) = \{(W,f): p \in W, W \textit{ open in } M, f: W \rightarrow R \textit{ is smooth}\}$$

Definition (Tangent vector): A tangent vector at $p \in M$ is a linear map $\xi: A(p) \rightarrow R$ that satisfies the following, all $f, g \in A(p)$:

1. If $f = g$ in a neighborhood of p, then $\xi(f) = \xi(g)$.
2. Product rule: $\xi(fg) = f(p)\xi(g) + \xi(f)g(p)$.

The tangent space to p at M, denoted by $T_p M$, is the set of all tangent vectors to M at p.

Definition (Vector field): A vector field on n-dimensional manifold M at p is a mapping $X: T_pM \rightarrow N$. A smooth vector field on a manifold is an assignment of $X_p \in T_p M$ for each $p \in M$, such that if $f: M \rightarrow R$ is a smooth function, then

$$(X,f) \equiv X_p(f): M \rightarrow R$$

is smooth over p.

Definition (Integral curve): Given a vector field f on M, a smooth curve $c:(t_1,t_2) \to M$ is called an *integral curve* if

$$\dot{c}(t) = f(c(t)), \textit{ for all } t \in (t_1,t_2)$$

An integral curve on M is a curve that follows a given vector field at each point. Vector fields thus represent differential equations on manifolds. The space of smooth vector fields becomes an algebra under the appropriate multiplication operation.

Definition (Lie algebra): A vector space V together with the binary operation $[\cdot, \cdot]$: $V \times V \to V$ is called a *Lie algebra* if $[\cdot, \cdot]$ satisfies the following for all A, B, $C \in V$ and r, $k \in R$:

1. Antisymmetry: $[A,B] = -[B,A]$.
2. Bilinearity: $[rA + kB, C] = r[A,C] + k[B,C]$.
3. $r[A,B] = [rA,B] = [A,rB]$.
4. Jacobi identity: $[A,[B,C]] + [B,[A,C]] + [C,[A,B]] = 0$.

The derivative of a smooth function λ in the direction of the vector field f is given by

$$L_f \lambda(p) = (f(p))(\lambda)$$

and is the definition of a Lie derivative.

Given two vector fields, f and g, a new vector field can be defined by

$$([f,g](p))\, \lambda = (L_f L_g \lambda)(p) - (L_g L_f \lambda)(p)$$

and $[f,g]$ is called the *Lie bracket* of f and g. A vector space with the Lie bracket as the binary operation forms a Lie algebra. Denote by $V^\infty(M)$ the space of C^∞ vector fields on M.

Together with the Lie bracket operation, $V^\infty(M)$ is an algebra.

Definition (Distribution): A distribution on M is a mapping Δ: $M \to TM$ that assigns to each $p \in M$ a subspace of the tangent space T_pM. Furthermore, Δ is called a *smooth distribution* if for every $p \in M$ there exists a neighborhood U and a collection of smooth vector fields f_i, $i = 1,\ldots,m$, such that for all $q \in U$

$$\Delta(q) = spam(f_i(q) |\ i = 1,\ldots,m$$

The dimension of $\Delta(q)$ is m.

A distribution Δ is called *involutive* if, given any two vector fields $f,g \in \Delta$, the Lie bracket (f,g) is also in Δ.

A distribution Δ defined on an open set U is called *nonsingular* if there exists some integer m such that for all $q \in U$

$$\dim\,(U\Delta\,(q)) = m$$

Definition (Lie group): A lie group is a group G that is also a differential manifold such that for x, $y \in G$,

1. $(x, y) \mapsto xy$
2. $x \mapsto x^{-1}$

are smooth functions.

All finite dimensional Lie groups may be represented as matrix groups. The matrix groups $GL(n,R)$, $O(n)$, $SO(n)$, $E(n)$, and $SE(n)$ are all Lie groups.

Let G be a Lie group with identity I, and let X_I be a tangent vector to G at I. For any $g \in G$, define the left translation by g to be a map $L_g : G \rightarrow G$ such that $L_g(x) = gx$, where $x \in G$. Since G is a Lie group, L_g is a diffeomorphism of G for each g. Diffeomorphism is a smooth inveritable function that maps one differential manifold to another. Taking the differential of L_g at e results in a map from the tangent space of G at e to the tangent space of G at g:

$$dL_g{:}T_gG \rightarrow T_gG$$

such that

$$X_g = dL_g(X_g)$$

The vector field formed by assigning $x_g \in T_g(G)$ for each $g \in G$ is called a *left invariant field*. The left invariant vector fields of G form an algebra under $[\cdot, \cdot]$, which is called the Lie algebra of G and denoted by $L(G)$. $L(G)$ is actually a subalgebra of the Lie algebra of all smooth vector fields on G.

The mathematical framework presented here will be utilized for developing the vehicle kinematics and for studying controllability of the kinematic model of autonomous underwater vehicles. The vehicle kinematics are obtained using rotations and translations, which are represented as Lie groups in the general framework of manifolds. However, the control design is done in terms of local coordinates. These local coordinates are the parameterizations of the global coordinates, and therefore have singularities, which are also present in the control design.

2 Problem Formulation and Examples

This chapter gives an overview of the motion planning and issues related with the various motion planning tasks of the autonomous underwater vehicles. The control of the kinematic model obtained for such vehicles involves the concepts of nonholonomy. It will be seen that the vehicles are nonlinear and underactuated in nature because of the nonholonomic constraints on their generalized velocities. Finally, the motion planning problem will be formulated for autonomous underwater vehicles, and issues related with the various motion planning tasks and the feedback control design for these examples will be discussed.

2.1 MOTION PLANNING OF NONHOLONOMIC SYSTEMS

The initial motivation for the work presented here comes from the research work done in order to do the motion planning and control design for the nonholonomically constrained systems. Motion planning for nonholonomic systems has been studied in great detail, and a lot of research is being done in this field. This problem has attracted researchers because of its challenging theoretical nature and practical importance. The nonholonomic constraints arise in a number of advanced robotic systems, and the application of such systems is numerous. The problem is also interesting because its theoretical behavior presents a number of challenges. First, such systems are underactuated, that is, the number of control inputs is less than the number of the states or the variables of the system to be controlled. Thus, motion planning implies that the systems can be completely controlled with a fewer number of actuators, thereby improving the overall cost-effectiveness of the system. Also, underactuation can provide backup control techniques for a fully actuated system. Second, both planning and control are more difficult than for holonomic systems. Some of the motion control problems that have been studied in detail are those of regulation (stabilization) and tracking.

The problem of stabilizing such systems is a big issue, as it has been proved by Brockett [2] that the nonholonomic control systems with restricted mobility cannot be stabilized to a desired configuration (equilibrium) using a smooth time-invariant state feedback law. Because of this fact, there has been extensive study of this problem. Some authors have proposed nonsmooth or discontinuous control laws. Others have proposed smooth but time-varying control laws for the purpose of regulation, and some have proposed the combination of both, that is, discontinuous time-varying control laws [8, 9]. The method of transforming the kinematic model into the chained-form model and doing the control on the same was first proposed by [10] for the case of a car-like robot. The study of feedback control of a nonholonomic car-like robot is done in [1]. Various motion planning tasks, such as tracking a time-varying reference trajectory, path following, and point-to-point stabilization of a car-like robot, were presented in [1]. The work presented in Chapter 3 is along the same lines as [1], extended and modified for the application of underwater vehicles. The design of feedback controllers will be used for different motion tasks utilizing the kinematic model of the system. The kinematic model will be developed using the definition of nonholonomic constraints.

2.2 NONHOLONOMIC CONSTRAINTS

System constraints on the mechanical systems whose expression involves generalized coordinates and velocities are known as *kinematic constraints* of the system, which are also known as the *nonholonomic constraints*. These are of the following form

$$a_i(q,\dot{q}) = 0, \quad i = 1,2,\ldots,k < n \tag{2.1}$$

where q is the generalized coordinate vector or the state vector. $q \in M \subset R^n$, where n is the dimension of the configuration space M, to which the vector q belongs. These will limit the admissible motions of the system by restricting the set of generalized velocities that can be attained at a given configuration. Usually such constraints are in mechanics in the Pfaffian form:

$$a_i^T(q)\dot{q} = 0, \quad i = 1,2,\ldots,k < n \tag{2.2}$$

or

$$C(q)\dot{q} = 0 \tag{2.3}$$

This means they are linear in the generalized velocities. $a_i(q) \in M \subset R^n, i = 1,2,\ldots,k$, are row vectors. The vectors $a_i : M \mapsto R^n$ are assumed to be smooth and linearly independent. The matrix $C(q) \in R^{n \times n}$ is a constraint matrix.

The kinematic constraints restrict the motion by limiting the set of generalized velocities. The nonholonomic constraints cannot be integrated to the positions. Thus, while the instantaneous mobility that a system can perform is restricted to $(n - 1)$-dimensional null space of the constraint matrix $C(q)$, we can still say that it is possible that any configuration in the state space M can be reached. In general, for a system with n coordinates and k nonholonomic constraints, although the velocities are restricted to $(n - k)$-dimensional space, the global controllability in the configuration space is still attainable.

The kinematic model of an underwater vehicle or any such system is developed without taking into consideration the actual dynamic forces acting on it. Such systems have kinematic constraints on their velocities. These constraints mostly arise due to the rolling of two surfaces against one another and a roll without the slip condition, as in case of a wheel and the road. These can also arise due to the conservation laws applicable to the system or from the nature of the control inputs physically applied to the system [11]. Thus, nonholonomic constraints allow the global movement of the system in the configuration space, while at the same time restricting or reducing the degrees of freedom or motion performed locally by the system.

The concept of nonholonomy is related to controllability of the corresponding control system. Redefining the constraint specification as the directions or degrees of freedom in which the system can move, rather than the direction in which it cannot move, is equivalent to stating the controllability problem of the corresponding control system. Thus, we can safely say that if the system is maximally nonholonomic, the system is controllable, as any point in the configuration space can be reached. This way a motion problem can be converted into a control problem.

Nonholonomic constraints arise in a number of ways and in various mechanical systems and applications. These can arise because of the reasons already given in the previous paragraph. For a more detailed analysis, the reader is referred to [11] and [12]. Some of the typical examples of the nonholonomic systems can be summarized as

- Wheeled mobile robots
- Space robots
- Underwater vehicles
- Satellites
- Multifingered hand manipulators
- Hopping robots

2.3 PROBLEM DESCRIPTION

The motion planning tasks for nonholonomic systems are attained through the use of the feedback controllers. The basic motion tasks considered for an underwater vehicle are as follows:

Point-to-point motion: Here a desired goal configuration must be reached by a system starting from a given initial configuration.

Path following: Here the system has to reach a desired final configuration starting from a given initial configuration, while at the same time it has to follow a given geometric path. The initial configuration can be considered to be either on or off the path.

Trajectory following: Here the system must reach a final configuration while following a trajectory in the Cartesian space (i.e., a geometric path with an associated timing law) starting from a given initial configuration (either on or off the trajectory).

The tasks are assumed here such that the systems work in an obstacle-free environment and are shown in Figure 2.1a to c for a car-like robot.

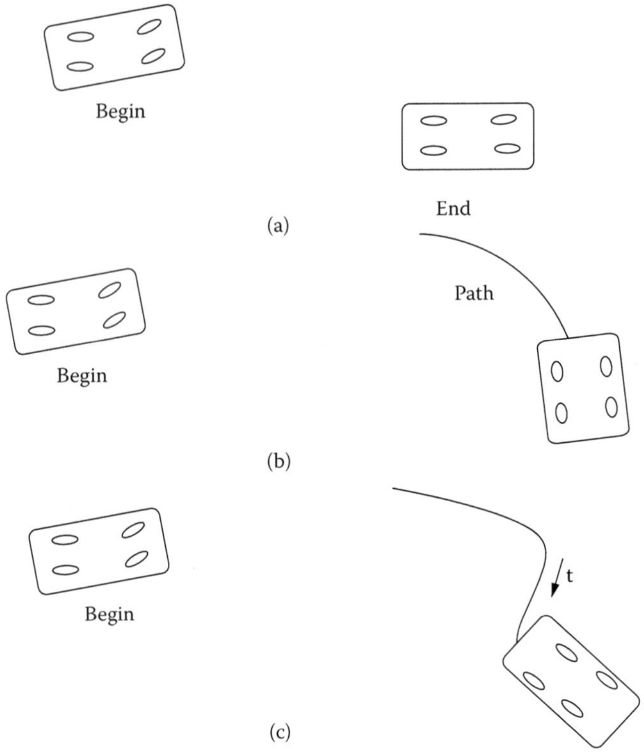

FIGURE 2.1 Motion planning tasks for a car-like robot.

The tasks can be obtained using either the feed-forward (open-loop) or feedback (closed-loop) control, or a combination of both. Since the feedback control is generally robust and can work well in the presence of disturbances, use of it is preferred.

Thinking in terms of controls, a point-to-point task can be thought of as a regulation control problem or a posture stabilization problem for an equilibrium point in the state space. The trajectory following is a tracking problem such that the error between the reference and the desired trajectories asymptotically goes to zero.

For nonholonomic systems, the tracking or path following, or both, is easier than the stabilization, whereas usually the reverse is true. This difference can be explained by drawing a comparison between the numbers of inputs and outputs (or states) to be controlled. In case of a regulation problem, m inputs (two in case of a car-like robot) are required to regulate or control n independent control variables or states (four in the case of a car-like robot), with m less than n. Thus, point-to-point stabilization is the most difficult of the three. In case of path following and trajectory tracking, the output to be controlled has a dimension (p) equal to that of the input (m). Thus, these control problems are square and their difficulty level is similar to and less than the stabilization one. For a car-like robot, in the case of path following m is 1 and p is 1, while for trajectory tracking m is 2 and p is 2; that is, we have to stabilize to zero the two-dimensional error vector associated with the Cartesian trajectory [1].

2.4 CONTROL MODEL FORMULATION

In this section we will be formulating the control model for the nonholonomic systems. For developing the control model we will be considering the first-order kinematic constraints on the system. As seen in Section 2.2, such constraints are of the following form:

$$a_i^T(q)\dot{q} = 0 \quad i = 1, 2, \ldots, k < n$$

or

$$C(q)\dot{q} = 0$$

where vector q is the generalized coordinates, $\dot{q}$ the first-order derivative (velocities) of the coordinates, and $C(q)$ the constraint matrix.

Let us denote the set of vector fields spanning the m-dimensional distribution Δ, which is annihilated by the constraints as g_j's, $j = 1, 2, \ldots, m$, such that

$$\Delta = span\{g_1, g_2, \ldots, g_m\}$$

The g_j's are the basis for the $n - k$ right null space of the constraint matrix $C(q)$ so that we have

$$a_i^T(q)g_j(q) = 0; \quad i = 1, 2, \ldots, k < n \quad j = 1, 2, \ldots, (n-k) = m \tag{2.4}$$

or

$$C(q)G(q) = 0 \tag{2.5}$$

The vector fields g_j's are assumed to be smooth and linearly independent as a consequence of the assumption on $a_i^T(q)'$s being smooth and independent. By expressing all the feasible velocities as a linear combination of these basis vectors, we obtain the first-order kinematic model of the system as

$$\dot{q} = g_1(q)v_1 + g_2(q)v_2 + \cdots + g_m(q)v_m \tag{2.6}$$

or

$$\dot{q} = \sum_{j=1}^{n-k=m} (g_j(q)v_j) \tag{2.7}$$

where v_j's, known as *pseudovelocities*, are taken as the control inputs and g_j's are the input vector fields. The model directly shows the presence of k nonholonomic constraints on the system having n states or configuration variables and $m = (n - k)$ control inputs. The control model in Equation 2.6 is known as the *kinematic model* of the system. The model is a driftless (i.e., no motion takes place under zero-input conditions), nonlinear, and underactuated (number of control inputs is less than the number of states to be controlled) control system.

2.5 CONTROLLABILITY ISSUES

Since the control model is driftless, the terms *local accessibility* and *controllability* can be used interchangeably. Moreover, the controllability of the whole configuration space is the (complete) nonholonomy of the kinematic constraints. The controllability condition can be established using Chow's theorem. According to the theorem, for driftless control systems, if the accessibility rank condition

$$\dim \Delta_c(q_0) = n \tag{2.8}$$

holds, then the control system is locally accessible (controllable) from q_0. Δ_c is the accessibility distribution of the kinematic model given by Equation 2.6 and is defined as the span of all the input vector fields

$$\Delta_c = span\{v | v \in \Delta_i \forall i \geq 1\}$$

with

$$\begin{aligned} \Delta_i &= \Delta_{i-1} + span\{[g, v] | \, g \in \Delta_1, v \in \Delta_{i-1}\}, \quad i \geq 2 \\ \Delta_1 &= span\{g_1, g_2, \ldots, g_m\} \end{aligned} \tag{2.9}$$

This implies that Δ_c is the involutive closure under Lie bracketing of the distribution Δ_1 associated with the input vector fields g_j's. The term $[g, v]$ is the Lie bracket of the two vector fields g and v, defined as

$$[g, v](q) = \frac{\partial v}{\partial q} g(q) - \frac{\partial g}{\partial q} v(q) \tag{2.10}$$

The Chow's theorem provides both a necessary and sufficient condition for the controllability [12]. Moreover, if the system is controllable, then its dynamic extension given by

$$\dot{q} = \sum_{j=1}^{n-k=m} (g_j(q) v_j) \tag{2.11}$$

with $\dot{v}_j = u_j$; $j = 1, 2, \ldots, m$, is also controllable. In some cases, the use of the nilpotent basis is made; that is, the input vector fields $g_j's$ are the nilpotent basis. This eliminates the need for cumbersome computations, as we will see that, using this concept, all higher-order Lie brackets above some particular order are zero [12].

2.6 STABILIZATION

The stabilization problem for the control system (Equation 2.7) can be defined as finding the feedback control law of the form $u(q,t)$ in order to make the closed-loop system asymptotically stable about an equilibrium point or a reference (feasible) trajectory. In the point stabilization problem, we assume an equilibrium point for the open-loop system, that is, $\dot{q} = f(q_e) = 0$.

2.6.1 Controllability and Stabilization at a Point

For the driftless control system (Equation 2.7), any configuration (q_e) is an open-loop equilibrium point under zero-input conditions. For linear systems $\dot{x} = Ax + Bu$, it is a well-established fact that if the system satisfies the controllability rank condition given by

$$rank[B\ AB\ A^2B\ \ldots\ A^nB] = n \tag{2.12}$$

then the asymptotic (exponential) stabilization by a smooth time-invariant state feedback is guaranteed. In other words, we can say that if the controllability condition is satisfied, there exists a feedback law $u = k(x - x_e)$, such that the closed-loop system is asymptotically stable about the equilibrium point x_e.

For the control model (Equation 2.7) we would like to make a similar kind of analysis. For this purpose, we will look at the approximate linearization of the system at any equilibrium point (q_e). The approximate linearization given by

$$\dot{q} = \delta\dot{q} = G(q_e)v \tag{2.13}$$

with $\delta q = q - q_e$, is clearly not controllable, as the rank of the controllability matrix $G(q_e)$ is m (which is less than n). Hence, we can say that a linear controller cannot achieve stabilization, not even locally.

However, the controllability of the nonlinear system can be established by using the tools from differential geometry, that is, we can make use of the Lie algebra rank condition to prove its controllability. However, even if the system can be proven to be globally controllable (in a nonlinear sense), there is still a severe theoretical limitation on the point stabilization. The limitation is in a sense that a Lyapunov (asymptotic) stability cannot be achieved by means of a smooth time-invariant feedback [13].

The above result can be established on the basis of Brockett's theorem [2], which says that the stabilization of a driftless regular system by a smooth time-invariant feedback is not possible. For the driftless, underactuated control system (Equation 2.7), where vector fields $g_j' s$ are linearly independent (regular) at q_e, the theorem implies the number of inputs m is equal to the number of states n as both a necessary and sufficient condition for smooth stabilization. Also, it should be noted that if the system cannot be stabilized by a smooth feedback, the same negative result is true for its dynamic extension, and also the theorem is not applicable to time-varying feedback laws.

Thus, in order to design the feedback controllers for posture stabilization, it is obligatory either to give up the continuity requirement, that is, include the nonsmooth feedback, or to apply the time-varying laws, or a combination of both.

2.6.2 Controllability and Stabilization about Trajectory

In case of trajectory following, for stabilization about a trajectory it is ensured that the reference trajectories are feasible for the system. In other words, we should take or generate only those state and input trajectories that satisfy the nonholonomic constraints of the system, that is, should satisfy Equation 2.2.

2.6.3 Approximate Linearization

For approximate linearization of the system (Equation 2.7) we take the desired state trajectory as $q_d(t)$ and the input trajectory as $v_d(t)$. It can be easily seen that the linearization about a smooth trajectory results in a linear time-varying system. The system can easily be shown to satisfy the controllability condition, that is, the controllability Grammian is nonsingular [21] as long as the input reference trajectory is persistent (does not come to a stop). Thus, it implies that we can achieve stabilization about the desired trajectory via a smooth time-invariant control law as long as the trajectories do not come to stop. One observation should be made: as we know, the control scheme presented here is based on the approximate linearization of the original system in the neighborhood of a reference trajectory; the closed-loop system is asymptotically stable only locally. In order to achieve global stabilization for trajectory tracking error, we have to make use of the nonlinear feedback design.

2.6.4 Exact Feedback Linearization

It is well known in robotics that if the number of generalized coordinates equals the number of control inputs, that is, $n = m$, the system kinematics or dynamics can be transformed into a linear system with the use of a nonlinear static state feedback [14]. The linearity is displayed by the system equations by performing some sort of coordinate transformation in the state space.

For exact linearization of the nonlinear systems, outputs are chosen to which a desired behavior is assigned. Two types of exact linearization are possible. The two schemes are full-state feedback linearization and input–output linearization. In the first case, the feedback transformation is such that the whole set of system equations become linear, while in the second case, the transformation is such that the input and output response of the closed-loop system is linear. For multi-input-multi-output (MIMO) systems this transformation results in decoupling of the input and output vectors. Both the transformations can be achieved through static feedback or dynamic feedback [14].

2.6.5 Static Feedback Linearization

For the nonholonomic kinematic model the full-state feedback linearization cannot be achieved using a static (time-invariant) state feedback. The reason for this is the violation of the necessary condition for the full-state feedback linearizability according to [14]. The fact is that the controllability condition for the system derived in Section 2.6 requires that the distribution Δ_o generated by g_j is not involutive, which violates the necessary condition for static feedback transformation.

However, the input–output linearization is possible with the use of static feedback. Here m equations are transformed via the feedback into simple decoupled integrators. However, the choice of outputs that are linearized is not unique. Here it is worth noticing that in the case of (2, n) chained-form transformation, the two variables are indeed examples of linearizing outputs with static feedback given by the input transformation equation. Also, it must be noted that in the case of input–output linearization the internal dynamics may be left in the closed-loop system. Thus, for the exponential (global) convergence of the trajectory error to go to zero, these internal dynamics should be properly modeled, analyzed, and their stability guaranteed.

2.6.6 Dynamic Feedback Linearization

For the exact feedback linearization, if the static feedback design fails, we can also make use of the dynamic control feedback design for nonholonomic systems. The use of dynamic feedback can also result in full-state feedback linearization. For the model in Equation 2.7,

$$\dot{q} = G(q)v; \quad q \in R^n, \quad v \in R^m \tag{2.14}$$

the dynamic feedback compensator is of the form

$$\begin{aligned} v &= a(q,\zeta) + b(q,\zeta)r \\ \dot{\zeta} &= c(q,\zeta) + d(q,\zeta)r \end{aligned} \tag{2.15}$$

where $\zeta(t) \in R^v$ is the compensator state vector of dimensions v, and $r(t) \in R^v$ (having the same dimensions as the input vector $\zeta(t)$) is the auxiliary input. Equation 2.15 is such that the closed-loop system obtained from Equations 2.14 and 2.15 is equivalent to a linear system under some state transformation $z = T(q,\zeta)$. For applications to the nonholonomic systems, the linearization process involves the following procedure.

Initially, we define the output of the system $y = h(q)$. To this output a desired behavior is assigned (track a trajectory). Then the output y is successively differentiated until the system inputs appear explicitly in a nonsingular way. The nonsingularity is a must for the inversion of the differentiated equations to solve for the inputs. If in a step involving the differentiation of system outputs the decoupling matrix (differential map) of the system is *singular* (which means that some input is still not appearing), integrators are added on some of the input channels and the process of differentiation is continued. It is also necessary to avoid the direct differentiation of the system inputs in the next differentiation. This operation is known as *dynamic extension* and converts a system input into a state of dynamic compensator. The dynamic compensator has the new auxiliary input r as its input. The process of differentiation continues until at some point the system is invertible (i.e., a solution for new inputs can be obtained) from the chosen output vector y and the process terminates. The number of successive addition of integrators gives dimensions of the state ζ of the dynamic compensator. Also, if the sum of the orders of the output differentiation is equal to the dimensions of the extended state space system (original and dynamic compensator state), which is $n + v$, then the full-state linearization of the system is obtained, as there are no internal dynamics left in the system.

2.7 EXAMPLES OF NONHOLONOMIC SYSTEMS

The simplest example of a nonholonomic system can be a wheel that rolls on a lane surface, such as a unicycle. The constraints here arise due to the roll without a slip condition. The configuration or the generalized coordinate vector is $q = (x, y, \theta)$. The coordinates x and y are the position coordinates of the wheel, and θ is the angle that the wheel makes with the x axis. The unicycle is shown in Figure 2.2. The constraint here is that the wheel cannot slip in the lateral direction.

The generalized velocities are subject to the following kinematic constraint:

$$\dot{x}\sin\theta - \dot{y}\cos\theta = 0 \tag{2.16}$$

In other words, the velocity along the plane perpendicular to the point of contact between the wheel and the ground is zero. The above equation is of the form $C(q)\dot{q} = 0$, with constraint matrix $C(q) = [\sin\theta \; -\cos\theta \; 0]$.

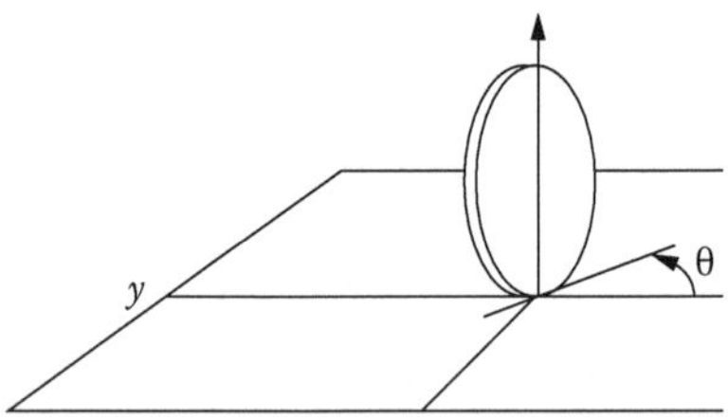

FIGURE 2.2 The nonholonomic constraints on a unicycle.

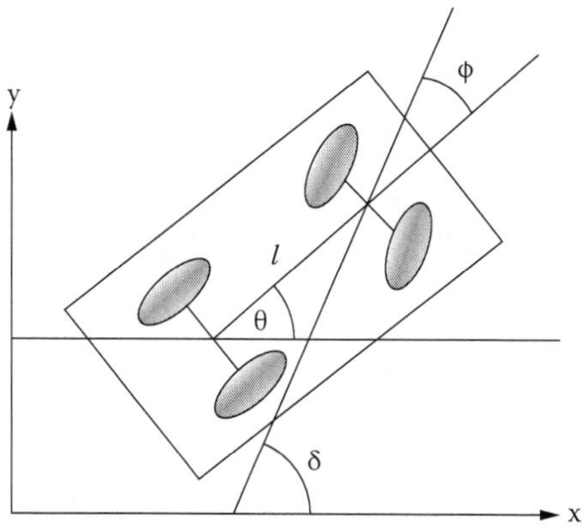

FIGURE 2.3 The nonholonomic constraints on a car-like robot.

Expressing the feasible velocities as a linear combination of vector fields spanning the null space of the matrix $C(q)$, we get the following kinematic model:

$$\dot{q} = g_1(q)v_1 + g_2(q)v_2$$

or

$$\begin{pmatrix} \dot{x} \\ \dot{y} \\ \dot{\theta} \end{pmatrix} = \begin{pmatrix} \cos\theta \\ \sin\theta \\ 0 \end{pmatrix} v_1 + \begin{pmatrix} 0 \\ 0 \\ 1 \end{pmatrix} v_2 \tag{2.17}$$

where v_1 is the linear velocity of the wheel and v_2 is its angular velocity around the vertical axis. Here we observe that the number of states $n = 3$, number of control inputs $m = 2$, and number of nonholonomic constraints $k = 1$.

Another example is that of a car-like robot shown in Figure 2.3. The robot has two wheels, and each wheel is subject to one nonholonomic constraint. The constraint is the same as in the case of a unicycle. The generalized coordinate vector is $q = (x, y, \theta, \phi)$, with x, y, and θ the same as before. The angle ϕ is the steering angle.

The two nonholonomic constraints on the front and rear wheels, respectively, are

$$\begin{aligned} \dot{x}\sin(\theta+\phi) - \dot{y}\cos(\theta+\phi) - \dot{\theta}l\cos\phi = 0 \\ \dot{x}\sin\theta - \dot{y}\cos\theta = 0 \end{aligned} \tag{2.18}$$

Here l is the distance between the wheels. Again, this is of the form $C(q)\dot{q} = 0$, with

$$C(q) = \begin{bmatrix} \sin(\theta+\phi) & -\cos(\theta+\phi) & -l\cos\phi & 0 \\ \sin\theta & -\cos\theta & 0 & 0 \end{bmatrix}$$

Choosing rear wheel drive, the kinematic model is obtained as

$$\dot{q} = g_1(q)v_1 + g_2(q)v_2$$

$$\begin{pmatrix} \dot{x} \\ \dot{y} \\ \dot{\theta} \\ \dot{\phi} \end{pmatrix} = \begin{pmatrix} \cos\theta \\ \sin\theta \\ \tan\phi/l \\ 0 \end{pmatrix} v_1 + \begin{pmatrix} 0 \\ 0 \\ 0 \\ 1 \end{pmatrix} v2 \tag{2.19}$$

Here v_1 is the rear driving velocity input and v_2 is the steering velocity input. The above model is not defined at $\phi = \pi/2$, where g_1 is discontinuous. Physically, this corresponds to the car becoming jammed because of its front wheel being normal to the axis of the body.

The feedback control design, controllability analysis, and motion planning for all three motion tasks are done in [1]. For a detailed analysis, the reader is referred to the same. Another example of nonholonomic systems is that of an underwater vehicle, which is discussed in the next chapter. In Chapter 3 we will study the motion planning for the same, and controllability of the system is discussed and proved. We also present feedback control laws that give global stabilization of the vehicle about a desired trajectory and about a point.

3 Mathematical Modeling and Controllability Analysis

In this chapter an overview of an underwater vehicle is given and a mathematical model of the vehicle is derived. The chapter presents in detail the derivation of the mathematical modeling of the system. For motion planning tasks the kinematic model of the system is obtained and the issues related with nonlinear controllability of the system are studied in detail. Finally, for the purpose of control design, the system is converted into a chained form.

3.1 MATHEMATICAL MODELING

In this section the mathematical model of an underwater vehicle is briefly discussed. An underwater vehicle is generally a six degrees of freedom body. It follows the laws of rigid body motion. The dynamics of the system are highly nonlinear due to rigid body coupling and hydrodynamic forces on the vehicle. The mathematical model of the underwater vehicle is obtained through the following two models:

Dynamic model: This type of model allows for the actual forces, causing the motion and the dynamic properties of the vehicle to be taken into account. The equations of translation and rotation are obtained using Newton's law [15].

Kinematic model: The kinematic model of the system is where the actual forces causing the motion and the dynamic properties of the vehicle do not enter the equations of motion. This type of model allows for the decoupling of the vehicle dynamics from its movement. An autonomous underwater vehicle has a nonholonomic nature due to its nonlinear kinematic model [22]. In the following section the kinematic model of the vehicle is derived. For the rest of this chapter and Chapter 4, the kinematic model of the system will be used for analysis and control purposes. Chapter 5 will introduce the dynamic model.

3.1.1 Kinematic Modeling and Nonholonomic Constraints

The kinematic model of the system is obtained by taking into consideration the nonholonomic constraints on the linear velocity. The nonholonomic constraints restrict the velocity of the system to be zero in certain directions, but these restrictions do not restrict the global movement of the system. Such constraints can arise when two surfaces roll against each other or in space-based systems where the total angular

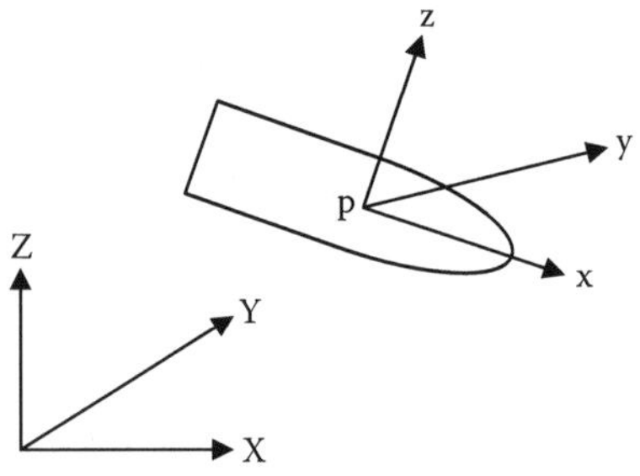

FIGURE 3.1 The coordinate systems of an underwater vehicle.

momentum of the system is conserved. For the development of the kinematic model of the underwater vehicle, we assume two orthogonal coordinate systems [23]:

Global coordinates: The global or the inertial frame coordinates are denoted by (O, X, Y, Z). The frame remains fixed at the ocean surface with origin O. The unit vector in the Z direction points down into the water, while the unit vectors along the X and Y direction complete the right-handed system.

Local coordinates: The local or the body frame coordinates are denoted by (p, x, y, z). The frame remains fixed on the vehicle surface with origin p. The two coordinate systems are as shown in Figure 3.1.

3.1.2 Kinematic Model with Respect to Global Coordinates

The kinematics of the vehicle is described by six state variables and four input variables. The kinematic relationships describing the transformations between the two coordinate systems can have a number of parameterizations. The one used here is the Euler angle parameterization [25]. In the Euler angle representation the orientation between the inertial and the local coordinate frame is expressed in terms of a sequence of three rotations, roll ϕ, pitch θ, and yaw ψ, about the axes x, y, and z, respectively.

Let $q = [p\ \eta]^T$ be the vector of six generalized coordinates required to specify the kinematics of the vehicle. The six coordinates are the Cartesian coordinate vector $p = [x, y, z]^T$ of the vehicle in the local frame and the orientation coordinate vector $\eta = [\phi, \theta, \psi]^T$. The orientation vector is the vector of Euler angles that gives the orientation of the body frame with respect to the inertial frame. The transformation from the local coordinate frame to the global coordinate frame is given by means of a transformation matrix $R \in S(O3)$ called *rotation matrix*, where $S(O3)$ is a group of rigid body rotations. The matrix R is an orthogonal matrix satisfying the relation $RR^T = I$, that is, $R^T = R^{-1}$ and $\det(R) = 1$. The matrix R is given as $R = [n\ s\ a]$ with the column vectors being

$$R = \begin{bmatrix} r_{11} & r_{12} & r_{13} \\ r_{21} & r_{22} & r_{23} \\ r_{31} & r_{32} & r_{33} \end{bmatrix} = [n \quad s \quad a] \tag{3.1}$$

with the matrix entries being

$$
\begin{aligned}
r_{11} &= \cos\theta\cos\psi \\
r_{12} &= \sin\theta\sin\phi\cos\psi - \cos\phi\sin\psi \\
r_{13} &= \sin\theta\cos\phi\cos\psi \\
r_{21} &= \cos\theta\sin\psi \\
r_{22} &= \sin\theta\sin\phi\sin\psi + \cos\phi\cos\psi \\
r_{23} &= \sin\theta\cos\phi\sin\psi - \sin\phi\cos\psi \\
r_{31} &= -\sin\theta \\
r_{32} &= \sin\phi\ \cos\theta \\
r_{33} &= \cos\phi\ \cos\theta
\end{aligned}
\tag{3.2}
$$

Let $v = [v_x \quad 0 \quad 0]^T$ be the linear velocity of the vehicle, that is, the vehicle has linear velocity along the x axis only, and let $\omega = [\omega_x \quad \omega_y \quad \omega_z]^T$ be the angular velocity components along the axes x, y, and z, respectively in the body frame. The velocity vector along the three coordinate axes and the time derivative of the Euler angles are obtained from the following relations:

$$\dot{p} = Rv = [n\ s\ a]v \tag{3.3}$$

$$\dot{R} = RS(\omega) \tag{3.4}$$

where $S(\omega)$ is the skew-symmetric matrix given as

$$S(\omega) = \begin{bmatrix} 0 & -\omega_z & \omega_y \\ \omega_z & 0 & -\omega_x \\ -\omega_y & \omega_x & 0 \end{bmatrix}$$

The above equations give the following on solving

$$\dot{p} = J_1(\eta)v,\ \dot{\eta} = J_2(\eta)\omega \tag{3.5}$$

with

$$J_1(\eta) = [\cos\theta\cos\psi \quad \cos\theta\sin\psi \quad -\sin\theta]^T$$

$$J_2(\eta) = \begin{bmatrix} 1 & \sin\phi\tan\theta & \cos\phi\tan\theta \\ 0 & \cos\phi & -\sin\phi \\ 0 & \sin\phi\sec\theta & \cos\phi\sec\theta \end{bmatrix}$$

The above set of equations can be written as the following equations:

$$\begin{aligned}
\dot{x}(t) &= r_{11}v = \cos\psi\cos\theta v \\
\dot{y}(t) &= r_{21}v = \sin\psi\cos\theta v \\
\dot{z}(t) &= r_{31}v = -\sin\theta v \\
\dot{\phi}(t) &= \omega_x + \sin\phi\tan\theta\omega_y + \cos\phi\tan\theta\omega_z \\
\dot{\theta}(t) &= \cos\phi\omega_y - \sin\phi\omega_z \\
\dot{\psi}(t) &= \sin\phi\sec\theta\omega_y + \cos\phi\sec\theta\omega_z
\end{aligned} \tag{3.6}$$

This can be written in matrix form as

$$\begin{bmatrix} \dot{x}(t) \\ \dot{y}(t) \\ \dot{z}(t) \\ \dot{\phi}(t) \\ \dot{\theta}(t) \\ \dot{\psi}(t) \end{bmatrix} = \begin{bmatrix} \cos\psi\cos\theta & 0 & 0 & 0 \\ \sin\psi\cos\theta & 0 & 0 & 0 \\ -\sin\theta & 0 & 0 & 0 \\ 0 & 1 & \sin\phi\tan\theta & \cos\phi\tan\theta \\ 0 & 0 & \cos\phi & -\sin\phi \\ 0 & 0 & \sin\phi\sec\theta & \cos\phi\sec\theta \end{bmatrix} \begin{bmatrix} v \\ \omega_x \\ \omega_y \\ \omega_z \end{bmatrix} \tag{3.7}$$

The equations can be written in generalized vector form as

$$\begin{bmatrix} \dot{x}(t) \\ \dot{y}(t) \\ \dot{z}(t) \\ \dot{\phi}(t) \\ \dot{\theta}(t) \\ \dot{\psi}(t) \end{bmatrix} = \begin{bmatrix} \cos\psi\cos\theta \\ \sin\psi\cos\theta \\ -\sin\theta \\ 0 \\ 0 \\ 0 \end{bmatrix} v + \begin{bmatrix} 0 \\ 0 \\ 0 \\ 1 \\ 0 \\ 0 \end{bmatrix} \omega_x + \begin{bmatrix} 0 \\ 0 \\ 0 \\ \sin\phi\tan\theta \\ \cos\phi \\ \sin\phi\sec\theta \end{bmatrix} \omega_y + \begin{bmatrix} 0 \\ 0 \\ 0 \\ \cos\phi\tan\theta \\ -\sin\phi \\ \cos\phi\sec\theta \end{bmatrix} \omega_z \tag{3.8}$$

The system here is subject to two nonholonomic constraints. The constraints are on the linear velocities along the y and z directions. The velocities along these directions are zero. The two constraints are as

$$\begin{aligned}
s^T\dot{p} &= 0 \\
a^T\dot{p} &= 0
\end{aligned} \tag{3.9}$$

which can be written as

$$\begin{aligned}
r_{12}\dot{x} + r_{22}\dot{y} + r_{32}\dot{z} &= 0 \\
r_{13}\dot{x} + r_{23}\dot{y} + r_{33}\dot{z} &= 0
\end{aligned}$$

Alternately, these equations can be written as

$$(\cos\psi\sin\theta\sin\phi - \sin\psi\cos\phi)\dot{x} + (\sin\psi\sin\theta\sin\phi + \cos\psi\cos\phi)\dot{y} + (\cos\theta\sin\phi)\dot{z} = 0$$

$$(\sin\psi\sin\theta\cos\phi - \sin\psi\sin\phi)\dot{x} + (\sin\psi\sin\theta\cos\phi - \cos\psi\sin\phi)\dot{y} + (\cos\theta\cos\phi)\dot{z} = 0$$

The above equations are of the form $A(q)\dot{q} = 0$, with

$$A(q) = \begin{bmatrix} r_{12} & r_{22} & r_{32} & 0 & 0 & 0 \\ r_{13} & r_{23} & r_{33} & 0 & 0 & 0 \end{bmatrix} \tag{3.10}$$

Expressing the feasible velocities as the linear combination of vector fields $g_1(q)$, $g_2(q)$, $g_3(q)$, and $g_4(q)$ spanning the null space of matrix $A(q)$, we have the following kinematic model:

$$\dot{q}(t) = g_1(q)v_1 + g_2(q)v_2 + g_3(q)v_3 + g_4(q)v_4$$

$$\dot{q}(t) = [g_1(q) \quad g_2(q) \quad g_3(q) \quad g_4(q)] \begin{bmatrix} v_1 \\ v_2 \\ v_3 \\ v_4 \end{bmatrix} \tag{3.11}$$

with $v_1 = v_x; v_2 = \omega_x; v_3 = \omega_y; v_4 = \omega_z$ and

$$\begin{aligned} g_1(q) &= [\cos\theta\cos\psi \quad \cos\theta\sin\psi \quad -\sin\theta \quad 0 \quad 0 \quad 0]^T \\ g_2(q) &= [0 \quad 0 \quad 0 \quad 1 \quad 0 \quad 0]^T \\ g_3(q) &= [0 \quad 0 \quad 0 \quad \sin\phi\tan\theta \quad \cos\phi \quad \sin\phi\sec\theta]^T \\ g_4(q) &= [0 \quad 0 \quad 0 \quad \cos\phi\tan\theta \quad -\sin\phi \quad \cos\phi\sec\theta]^T \end{aligned} \tag{3.12}$$

More generally we can write

$$\dot{q} = G(q)v \tag{3.13}$$

The above equations are the kinematic model of the system. The system is nonlinear and underactuated, which means that the number of inputs to the system is less than its states. The generalized velocity vector $\dot{q}$ cannot assume any independent value unless it satisfies the nonholonomic constraints. The constraints are examples of the Pfaffian constraints, which are linear in velocities. The admissible generalized velocities as given by Equation 3.1 are contained in the null space of the constraint matrix $A(q)$.

3.2 CONTROLLABILITY ANALYSIS

Consider the system in Equation 3.13,

$$\dot{q} = G(q)v$$

The system is nonlinear, underactuated, and driftless; that is, no motion takes place under zero-input conditions. Thus, in order to establish the controllability of the vehicle, we make use of the mathematical concepts that are involved in the Lie algebra rank condition and differential geometric control theory concepts.

3.2.1 Controllability about a Point

Consider the linear approximation of the system (Equation [3.13]) at equilibrium point q_e while letting the input v_e equal zero. Let the error associated with the equilibrium point be given as

$$\tilde{q} = q - q_e$$

The time derivative of the error is given as

$$\dot{\tilde{q}} = g_1(q_e)v_1 + g_2(q_e)v_2 + g_3(q_e)v_3 + g_4(q_e)v_4$$

or

$$\dot{\tilde{q}} = G(q_e)v \tag{3.14}$$

Here $G(q_e)$ is the controllability matrix at the equilibrium point. The rank of the controllability matrix is 4. Thus, if we linearize the system about an equilibrium point, the linearized system is not controllable. Hence, the linear controller will not work here. To test the controllability of the above system, we make use of the Lie algebra rank condition and nilpotent basis:

Nilpotent basis: The definition of nilpotent basis for a distribution is recalled here [16]. Given a set of generators or basis vector fields $g_1, g_2, g_3 \dots, g_m$, we define the length of a Lie product recursively as

$$l\{g_i\} = 1, \quad i = 1, 2, \dots, m$$

$$l\{[A, B]\} = l[A] + l[B]$$

where A and B are themselves Lie products. Alternatively, $l[A]$ is the number of generators in the expansion for A. The Lie algebra or basis is nilpotent if there exists an integer k such that all Lie products of length greater than k are zero. The integer k is called the *order of nilpotency* [17]. The use of the nilpotent basis eliminates the need for cumbersome computations, as we see that all higher-order Lie brackets above some particular order are zero.

In the light of the above definition and conditions, we see that the Lie algebra $L\{g_1, g_2, g_3, g_4\}$ is a nilpotent algebra of order 2 ($k = 2$); that is, the vector fields g_1, g_2, g_3, and g_4 are the nilpotent basis. Thus, all Lie brackets of order more than two are zero. The only independent Lie brackets computed from the four basis vector fields are $[g_1, g_3]$ and $[g_1, g_4]$.

Thus, for our system the Lie algebra rank condition becomes

$$rank[C_c] = 6$$

or

$$rank[g_1, g_2, g_3, g_4, [g_1, g_3][g_1, g_4]] = 6 \tag{3.15}$$

where $[g_1, g_3]$ and $[g_1, g_4]$ are two independent Lie brackets computed from the four vector fields $[g_1, g_2, g_3, g_4]$ per the following definition:

$$[g, h](x) = \frac{\partial h}{\partial x} g - \frac{\partial g}{\partial x} h \tag{3.16}$$

Thus, we have

$$[g_1, g_3](x) = \frac{\partial g_3}{\partial x} g_1 - \frac{\partial g_1}{\partial x} g_3$$

$$= [\cos\psi \sin\theta \cos\phi + \sin\psi \sin\phi \quad \sin\psi \sin\theta \cos\phi - \cos\psi \sin\phi \quad \ldots$$

$$\cos\theta \cos\phi \quad 0 \quad 0 \quad 0]^T$$

$$[g_1, g_4](x) = \frac{\partial g_4}{\partial x} g_1 - \frac{\partial g_1}{\partial x} g_4$$

$$= [-\cos\psi \sin\theta \sin\phi + \sin\psi \cos\phi \quad -\sin\psi \sin\theta \sin\phi - \cos\psi \cos\phi \quad \ldots$$

$$-\cos\theta \sin\phi \quad 0 \quad 0 \quad 0]^T$$

Using the above expressions for the Lie brackets, the controllability matrix C_c becomes

$$C_c = \begin{bmatrix} \cos\psi\cos\theta & 0 & 0 & 0 & \cos\psi\sin\theta\cos\phi + \sin\psi\sin\phi & -\cos\psi\sin\theta\sin\phi + \sin\psi\cos\phi \\ \sin\psi\cos\theta & 0 & 0 & 0 & \sin\psi\sin\theta\cos\phi - \cos\psi\sin\phi & -\sin\psi\sin\theta\sin\phi - \cos\psi\cos\phi \\ -\sin\theta & 0 & 0 & 0 & \cos\theta\cos\phi & -\cos\theta\sin\phi \\ 0 & 1 & \sin\phi\tan\theta & \cos\phi\tan\theta & 0 & 0 \\ 0 & 0 & \cos\phi & -\sin\phi & 0 & 0 \\ 0 & 0 & \sin\phi\sec\theta & \cos\phi\sec\theta & 0 & 0 \end{bmatrix}$$

The controllability matrix C_c has one nonzero minor of order six. Thus, the rank of the controllability matrix is full as long as $\theta \neq \pi/2$, which is the singularity of the system. Hence, we conclude that the system is controllable locally and also globally as long as it avoids the singularity condition.

3.2.2 Controllability about a Trajectory

For the nonlinear system

$$\dot{q} = G(q)v$$

let the reference state trajectory be $q_d(t) = [x_d(t), y_d(t), z_d(t), \phi_d(t), \theta_d(t), \psi_d(t)]^T$ and the reference input trajectory be $v_d(t) = [v_{d1}(t), v_{d2}(t), v_{d3}(t), v_{d4}(t)]^T$. The reference trajectory should satisfy the nonholonomic constraints on the system.

For the linear systems $\dot{x}(t) = Ax(t) + Bu(t)$ controllability implies asymptotic (actually exponential) stability by smooth state feedback. Thus, if the accessibility rank condition $rank\,[B, AB, A^2B, \ldots, A^{n-1}B] = n$ is satisfied, then there exists a feedback gain so that the control law

$$u(t) = k(x_d - x)$$

makes the desired trajectory $x_d(t)$ asymptotically stable, or in other words, the error associated with the desired solution goes exponentially to zero.

For nonlinear systems the condition does not apply as such. But for local accessibility we may look at the approximate linearization of the system in the neighborhood of $x_d(t)$. Thus, in particular, if the linearized system is controllable, the nonlinear system can be stabilized locally at $x_d(t)$ by a smooth feedback $u(t) = kx_e$. The condition is sufficient but not necessary.

Let the errors associated with the desired state trajectory and input trajectory be denoted as $q_e(t) = q(t) - q_d(t)$ and $v_e(t) = v(t) - v_d(t)$, respectively. Linearizing the system about the desired trajectory we obtain the following system:

$$\begin{aligned} \dot{q}(t) &= \dot{q}_d(t) + \dot{q}_e(t) \\ &= \{G(q_d + q_e, t)\}\{v_d(t) + v_e(t)\} \end{aligned} \tag{3.17}$$

The Taylor series expansion of $G(q, t)$ about the nominal solution $q_d(t)$ is given as

$$\dot{q}(t) = \{G(q_d, t) + (\partial G(q)/\partial q\,|_{q=q_d})q_e(t) + h.o.t.\}\{v_d(t) + v_e(t)\}$$

Since the nominal solution satisfies Equation 3.16, we have

$$\dot{q}_e(t) = \{\partial G(q)/\partial q\,|_{q=q_d}\, q_e(t)\}v_d(t) + G(q_d, t)v_e(t)$$

or

$$\dot{q}_e(t) = A(t)q_e(t) + B(t)v_e(t)$$

where

$$A(t) = \sum_{i=1}^{4} (\partial g_i/\partial q\,|_{q=q_d})v_{di}(t)$$

and

$$B(t) = G(q_d, t) \tag{3.18}$$

Upon computations we get

$$A(t) = \begin{bmatrix} O_{3\times 3} & A_1(t) \\ O_{3\times 3} & A_2(t) \end{bmatrix}$$

with

$$A_1(t) = \begin{bmatrix} 0 & -\cos\psi_d \sin\theta_d v_{d1}(t) & -\sin\psi_d \cos\theta_d v_{d1}(t) \\ 0 & -\sin\psi_d \sin\theta_d v_{d1}(t) & \cos\psi_d \cos\theta_d v_{d1}(t) \\ 0 & \cos\theta_d v_{d1}(t) & 0 \end{bmatrix}$$

and

$$A_2(t) = \left[\begin{matrix} \cos\phi_d \tan\theta_d v_{d3}(t) - \sin\phi_d \tan\theta_d v_{d4}(t) \\ -\sin\phi_d v_{d3}(t) - \cos\phi_d v_{d4}(t) \\ -\cos\phi_d \sec\theta_d v_{d3}(t) - \sin\phi_d \sec\theta_d v_{d4}(t) \end{matrix} \quad \cdots \right.$$

$$\left. \begin{matrix} \sin\phi_d \sec^2\theta_d v_{d3}(t) + \cos\phi_d \sec^2\theta_d v_{d4}(t) & 0 \\ 0 & 0 \\ \sin\phi_d \sec\theta_d \tan\theta_d v_{d3}(t) + \cos\phi_d \sec\theta_d \tan\theta_d v_{d4}(t) & 0 \end{matrix} \right]$$

and

$$B(t) = \begin{bmatrix} J_1(t) & O_{3\times 3} \\ O_{3\times 1} & J_2(t) \end{bmatrix}$$

where

$$J_1(t) = \begin{bmatrix} \cos\psi_d \cos\theta_d \\ \sin\psi_d \cos\theta_d \\ -\sin\theta_d \end{bmatrix}$$

and

$$J_2(t) = \begin{bmatrix} 1 & \sin\phi_d \tan\theta_d & \cos\phi_d \tan\theta_d \\ 0 & \cos\phi_d & -\sin\phi_d \\ 0 & \sin\phi_d \sec\theta_d & \cos\phi_d \sec\theta_d \end{bmatrix}$$

The above system is linear time varying, but for a linear reference trajectory with constant velocities $v_{di}(t) = v_{dn}$ and $q_d(0) = q(0)$ the controllability condition becomes

$$rank\,[B, AB, A^2B, A^3B, A^4B, A^5B] = 6$$

Upon computations we see that the above matrix has a nonzero minor of order six as long as $v_{d1} = v_{d3} = v_{d4} \neq 4$ and $\theta_d \neq \pi/2$. Thus, the linearized system is controllable along a reference trajectory as long as the trajectory does not collapse to a point.

3.3 CHAINED FORMS

The chained-form systems were first introduced in [18]. The chained form introduced in [18] had one chain, that is, two input chained forms. The method for converting the multi-input drift-free nonholonomic systems into chained forms is given in [19] and [20]. The chained form thus obtained has more than one chain. The method presented in [19] for transforming is similar to the method of exact linearization of nonlinear systems with drift via state feedback, as presented in [14]. The same method will be applied for our system with some modifications.

The system

$$\dot{q}(t) = g_1(q)v_1 + g_2(q)v_2 + g_3(q)v_3 + g_4(q)v_4$$

can be transformed into the chained form by a feedback transformation. In the above equation, with g_i being smooth and linearly independent vector fields, there exists a feedback transformation $(\xi, \alpha, \eta, \gamma) = \Phi(q)$ and $v = \beta(q)$ that transforms the system into the following chained form [21]:

$$\begin{aligned} &\dot{\xi} = u_1, \dot{\alpha}_0 = u_2, \dot{\eta}_0 = u_3, \dot{\gamma}_0 = u_3 \\ &\dot{\eta}_0 = \alpha_0 u_1, \dot{\eta}_1 = \eta_0 u_1 \end{aligned} \tag{3.19}$$

There exists a basis function f_1, f_2, f_3, f_4, for the distribution $\Delta_0 = span(g_1, g_2, g_3, g_4)$, having the form

$$
\begin{aligned}
f_1 &= \frac{\partial}{\partial q_1} + \sum_{i=2}^{6} f_i(q)\frac{\partial}{\partial q_i} \\
f_2 &= \sum_{i=2}^{6} f_2^i(q)\frac{\partial}{\partial q_i} \\
f_3 &= \sum_{i=2}^{6} f_3^i(q)\frac{\partial}{\partial q_i} \\
f_4 &= \sum_{i=2}^{6} f_4^i(q)\frac{\partial}{\partial q_i}
\end{aligned}
\tag{3.20}
$$

The basis functions are chosen such that the following distributions

$$
\begin{aligned}
G_0 &= span(f_2, f_3, f_4) \\
G_1 &= span(f_2, f_3, f_4, [f_1, f_2], [f_1, f_3], [f_1, f_4]) \\
&\dots \\
G_5 &= span\left(ad_{f_i}^i f_2, ad_{f_i}^i f_3, ad_{f_i}^i f_4\right);\ 0 \le i \le 5
\end{aligned}
$$

with

$$
ad_{f_i}^k f_2 := [f_1, ad_{f_1}^{k-1} f_2] \quad \text{and} \quad ad_{f_i}^0 f_2 := f_2
$$

have constant dimension on the same open set $U \in R^n$, are all involutive, and G_5 has dimension five on U.

The vector fields f_1, f_2, f_3, and f_4 which satisfy these conditions, are

$$
\begin{aligned}
f_1 &= \frac{g_1}{c\psi c\theta} = [1 \quad \tan\psi \quad -\tan\theta \sec\psi \quad 0 \quad 0 \quad 0]^T \\
f_2 &= g_2, f_3 = g_3, f_4 = g_4
\end{aligned}
\tag{3.21}
$$

The coordinate transformation for the system thus is

$$
\begin{aligned}
&\xi_0 = h_1, \alpha_0 = L_{f_1}^1 h_1, \eta_0 = L_{f_1}^0 h_3, \gamma_0 = h_4 \\
&\alpha_1 = L_{f_1}^0 h_2, \eta_1 = h_3
\end{aligned}
\tag{3.22}
$$

where h_1, h_2, h_3, h_4, are the smooth functions such that the following conditions are met:

$$dh_1 \perp G_j; 0 \le j \le 5$$

and the distribution G_0 is annihilated by dh_1, dh_2, $dL^0_{f_1}h_2$, $dL^1_{f_1}h_2$, dh_3, $dL^0_{f_1}h_3$, $dL^1_{f_1}h_3$, and dh_4. Here $L_{f_1}h_3$ is the Lie derivative of h_3 with respect to f_1. For the detailed proof of the above conditions, refer to [19]. Here it should also be noted that the choice of functions h_1, h_2, h_3, h_4 is not unique. Choosing

$$h_1 = x,\ h_2 = y,\ h_3 = z$$

and

$$h_4 = \frac{1}{1 + trace(R)}(r_{32} - r_{23})$$

with R being the rotation matrix and $trace(R) = (r_{11} + r_{22} + r_{33})$. Thus, the coordinate transformation for the system becomes

$$\begin{aligned}
x_1 &= \xi_0 = x \\
x_2 &= \alpha_0 = \tan\psi \\
x_3 &= \alpha_1 = y \\
x_4 &= \eta_0 = -\tan\theta\sec\psi \\
x_5 &= z \\
x_6 &= \frac{1}{1 + tr(R)}(r_{32} - r_{23})
\end{aligned} \tag{3.23}$$

which gives the following chained-form system:

$$\begin{aligned}
\dot{x}_1 &= u_1 \\
\dot{x}_2 &= u_2 \\
\dot{x}_3 &= x_2 u_1, \\
\dot{x}_3 &= x_2 u_1, \\
\dot{x}_4 &= u_3 \\
\dot{x}_5 &= x_4 u_1 \\
\dot{x}_6 &= u_4
\end{aligned} \tag{3.24}$$

Solving Equations 3.23 and 3.24 and inserting Equation 2.8, we get the input transformation as

$$u_1 = \dot{x}_1$$

$$u_2 = \dot{x}_2$$

$$u_3 = \dot{x}_4$$

$$u_4 = \dot{x}_6$$

or

$$u_1 = \cos\psi\cos\theta v_1$$

$$u_2 = \sec^2\psi\sin\phi\sec c\theta v_3 + \sec^2\psi\cos\phi\sec\theta v_4$$

$$u_3 = \frac{(-\sin\psi\sin\phi\sin\theta - \cos\psi\cos\phi)}{\cos^2\psi\cos^2\theta}v_3 - \frac{(\sin\psi\cos\phi\sin\theta + \cos\psi\sin\phi)}{\cos^2\psi\cos^2\theta}v_4$$

$$u_4 = \frac{(1+\cos\psi\cos\theta)v_2 + (\cos\psi\sin\theta\sin\phi - \sin\psi\cos\phi)v_3 + (\sin\psi\sin\theta\cos\phi + \sin\psi\sin\phi)v_4}{1+\cos\psi\cos\theta + \sin\psi\sin\theta\sin\phi + \cos\psi\cos\phi + \cos\theta\cos\phi} \tag{3.25}$$

which gives

$$v_x = v_1 = u_1/\cos\psi\cos\theta$$

$$\omega_y = v_3 = \cos\psi\cos\theta\{(\cos\psi\sin\phi - \sin\psi\sin\theta\cos\phi)u_2 - (\cos\phi\cos\theta)u_3\}$$

$$\omega_z = v_4 = \cos\psi\cos\theta\{(\sin\psi\sin\theta\sin\phi + \cos\psi\cos\phi)u_2 - (\sin\phi\cos\theta)u_3\}$$

$$\omega_x = v_2 = \frac{1}{\cos\psi\cos\theta}[(1+\cos\psi\cos\theta + \sin\psi\sin\theta\sin\phi + \cos\psi\cos\phi + \cos\theta\cos\phi)u_4 \ldots$$

$$-(\cos\psi s\theta\sin\phi - \sin\psi\cos\phi)v_3 - (\sin\psi\sin\theta\sin\phi + \sin\psi\sin\phi)v_4]$$

The inputs v_1, v_2, v_3, v_4 can be calculated from the above equations provided $\cos\psi\cos\theta \neq 0$. Also, here it should be noted that the chained-form system is completely controllable, as the controllability is not affected by state feedback and coordinate transformations, that is, they are invariant under the transformations. The chained form thus obtained is given as

$$\dot{x}_1 = x_1^0 = u_1,\ \dot{x}_2 = x_2^0 = u_2,\ \dot{x}_4 = x_3^0 = u_3,$$

$$\dot{x}_6 = x_4^0 = u_4,\ \dot{x}_3 = x_{20}^0 = x_2u_1,\ \dot{x}_5 = x_{31}^0 = x_4u_1$$

This form will be used for the control design, which is discussed in the next chapter.

4 Control Design Using the Kinematic Model

In this chapter, the controllers will be designed for the vehicle to track a desired trajectory, follow a path, and for point-to-point stabilization. The chapter presents the control design and the simulation results obtained for the model of an underwater vehicle developed in Chapter 3. The feedback control design is developed using the kinematic model of the system. The performance of the controllers obtained using various techniques of control design are evaluated for different motion planning tasks mentioned above. The chapter also presents the simulation results obtained for different controllers. The simulation results are used to compare and evaluate the performance of the various controllers.

4.1 TRAJECTORY TRACKING AND CONTROLLER DESIGN FOR THE CHAINED FORM

The system is supposed to track a given (desired) Cartesian trajectory. The problem is to regulate both the vehicle's position and orientation with respect to that of a reference system, the trajectory of which is parameterized by the variable t. The goal will be achieved using the feedback control law with the following control schemes:

- Full-state feedback using approximate linearization
- Feedback linearization using input–output linearization or full-state linearization

Before going for the feedback design, the problem of generating the desired output trajectory is discussed for both the original system and the chained-form system.

4.2 REFERENCE TRAJECTORY GENERATION

Let the reference state trajectory be $q_d(t) = [x_d(t), y_d(t), z_d(t), \phi_d(t), \theta_d(t), \psi_d(t)]^T$ and the reference input trajectory be $v_d(t) = [v_{d1}(t), v_{d2}(t), v_{d3}(t), v_{d4}(t)]^T$. The desired trajectory is feasible only when it satisfies the nonholonomic constraints on the system. Assume that a feasible and smooth desired output trajectory for the chained form is given as $[x_{d1}(t), x_{d3}(t), x_{d5}(t), x_{d6}(t)]^T$ $x_d(t)$. From this information we are able to derive the time evolution of the rest of the coordinates of the state trajectory and the associated input trajectory. In other words, we should be able to recover the state trajectory and the input trajectory from the reference output trajectory. From Equation 3.24 we have

$$\begin{aligned}
\dot{x}_{d1}(t) &= u_{d1}(t) \\
\dot{x}_{d2}(t) &= u_{2d}(t) \\
\dot{x}_{d3}(t) &= x_{d2}(t)u_{d1}(t) \\
\dot{x}_{d4}(t) &= u_{d3}(t) \\
\dot{x}_{d5}(t) &= x_{d4}(t)u_{d1}(t) \\
\dot{x}_{d6}(t) &= u_{d4}(t)
\end{aligned} \tag{4.1}$$

with initial conditions of the states as $[x_{d1}(t_0), x_{d2}(t_0), x_{d3}(t_0), x_{d4}(t_0), x_{d5}(t_0), x_{d6}(t_0)]^T$ at $t = t_0$. Solving for the state trajectory from Equation 4.1 we get

$$\begin{aligned}
x_{d2}(t) &= \dot{x}_{d3}(t)/\dot{x}_{d1}(t) \\
x_{d4}(t) &= \dot{x}_{d5}(t)/\dot{x}_{d1}(t)
\end{aligned} \tag{4.2}$$

The corresponding input trajectory is given as

$$\begin{aligned}
u_{d1}(t) &= \dot{x}_{d1}(t) \\
u_{d2}(t) &= \dot{x}_{d2}(t) = (\dot{x}_{d1}(t)\ddot{x}_{d3}(t) - \dot{x}_{d3}(t)\ddot{x}_{d1}(t))/\dot{x}_{d1}^2(t) \\
u_{d3}(t) &= \dot{x}_{d4}(t) = (\dot{x}_{d1}(t)\ddot{x}_{d5}(t) - \dot{x}_{d5}(t)\ddot{x}_{d1}(t))/\dot{x}_{d1}^2(t) \\
u_{d4}(t) &= \dot{x}_{d6}(t)
\end{aligned} \tag{4.3}$$

Equations 4.2 and 4.3 give the unique state and input trajectory from which the desired output trajectory can be reproduced or generated. As is seen, the values of the trajectories depend upon the values of the output trajectory and its second-order derivatives. Thus, the output trajectory should be differentiable everywhere. The derivation of the reference input and state trajectory, which generates a desired output trajectory, can also be performed on the original system. The original state and input trajectories can be derived from the output trajectory as

$$\begin{aligned}
x_d(t) &= x_{d1}(t) \\
y_d(t) &= x_{d3}(t) \\
z_d(t) &= x_{d5}(t) \\
\psi_d(t) &= \tan^{-1}(x_{d2}(t)) = \tan^{-1}(\dot{y}_d(t)/\dot{x}_d(t)) \\
\theta_d(t) &= -\tan^{-1}(x_{d4}(t)\cos\psi_d(t)) = -\tan^{-1}(\dot{z}_d(t)\cos\psi_d(t)/\dot{x}_d(t)) \\
\phi_d(t) &= \cot^{-1}(\cot\theta_d(t)/\sin\psi_d(t) + \tan\psi_d(t)/\sin\theta_d(t))
\end{aligned} \tag{4.4}$$

Similarly, the actual input trajectory is

$$
\begin{aligned}
v_{d1}(t) &= \sqrt{\dot{x}_d^2 + \dot{y}_d^2 + \dot{z}_d^2} = u_{d1}(t)/r_{d11}(t) \\
v_{d3}(t) &= -r_{d11}(t)(r_{d23}(t)u_{d2}(t) + r_{d33}(t)u_{d3}(t)) \\
v_{d4}(t) &= r_{d11}(t)(r_{d22}(t)u_{d2}(t) + r_{d32}(t)u_{d3}(t)) \\
v_{d2}(t) &= \frac{1}{1+r_{d11}(t)}(1 + r_{d11}(t) + r_{d22}(t) + r_{d33}(t))u_{d4}(t) - r_{d12}(t)v_{d3}(t) - r_{d13}(t)v_{d4}(t))
\end{aligned}
\tag{4.5}
$$

with

$$
\begin{aligned}
r_{d11} &= \cos\theta_d \cos\psi_d \\
r_{d12} &= \sin\theta_d \sin\phi_d \cos\psi_d - \cos\phi_d \sin\psi_d \\
r_{d13} &= \sin\theta_d \cos\phi_d \cos\psi_d + \sin\phi_d \sin\psi_d \\
r_{d21} &= \cos\theta_d \sin\psi_d \\
r_{d22} &= \sin\theta_d \sin\phi_d \sin\psi_d + \cos\phi_d \cos\psi_d \\
r_{d23} &= \sin\theta_d \cos\phi_d \sin\psi_d - \sin\phi_d \cos\psi_d \\
r_{d31} &= -\sin\theta_d \\
r_{d32} &= \sin\phi_d \cos\theta_d \\
r_{d33} &= \cos\phi_d \cos\theta_d
\end{aligned}
\tag{4.6}
$$

For the tracking simulation purposes consider the following reference sinusoidal output trajectory:

$$
x_{d1}(t) = t,\ x_{d3}(t) = A\sin\omega t,\ x_{d5}(t) = 1,\ x_{d6}(t) = 0 \tag{4.7}
$$

This gives the state trajectory as

$$
x_{d2}(t) = A\omega\cos\omega t,\ x_{d4}(t) = 0 \tag{4.8}
$$

and the input trajectory as

$$
\begin{aligned}
&u_{d1}(t) = 1,\ u_{d2}(t) = -A\omega^2 \sin\omega t \\
&u_{d3}(t) = 0,\ u_{d3}(t) = 0
\end{aligned}
\tag{4.9}
$$

The initial states are

$$x_{d1}(0) = 0 \quad x_{d2}(0) = A\omega \quad x_{d3}(0) = 0$$
$$x_{d4}(0) = 0 \quad x_{d5}(0) = 1 \quad x_{d6}(0) = 0 \tag{4.10}$$

Here again it is to be noted that there is a singularity in the state and input trajectories at $\dot{x}_{d1}(t) = 0$ or $u_{d1}(t) = 0$, as the state and input trajectories are not defined at that point.

4.3 CONTROL USING APPROXIMATE LINEARIZATION

The feedback controller for trajectory tracking is based on standard linear control theory. The design makes use of the approximate linearization of the system equations about desired trajectory, which leads to a time-varying system, as seen before. The method here is illustrated for the chained-form equations about the desired trajectory. The chained-form system is linear under piecewise constant inputs. For the chained-form system the desired state and input trajectory computed in correspondence to the reference Cartesian trajectory is

$$x_d(t) = \{x_{d1}(t), x_{d2}(t), x_{d3}(t), x_{d4}(t), x_{d5}(t), x_{d6}(t)\}$$

and

$$u_d(t) = \{u_{d1}(t), u_{d2}(t), u_{d3}(t), u_{d4}(t)\} \tag{4.11}$$

An equivalent way to state the tracking problem is to require the difference between the actual configuration and the desired configuration to approach to zero. This difference is denoted as the error. Since the vehicle will not necessarily share the same initial conditions as the desired system, the tracking controller will drive the error to zero and minimize the effect of the disturbances as the vehicle converges to the reference trajectory.

In order for the system to track the trajectory, this error should approach zero with time. Denoting the error variables for states and inputs as following

$$x_e = x - x_d$$

and

$$u_e = u - u_d \tag{4.12}$$

the error differential equations are written by subtracting the desired equations from the system (actual) equations as the following nonlinear set of equations:

$$\begin{aligned}
\dot{x}_{e1} &= u_{e1} \\
\dot{x}_{e2} &= u_{e2} \\
\dot{x}_{e3} &= x_2 u_1 - x_{d2} u_{d1} \\
\dot{x}_{e4} &= u_{e3} \\
\dot{x}_{e5} &= x_4 u_1 - x_{d4} u_{d1} \\
\dot{x}_{e6} &= u_{e4}
\end{aligned} \tag{4.13}$$

Now linearizing about the desired trajectory, we have the following linear system:

$$\dot{x}_e(t) = A(t)x_e(t) + B(t)u_e(t) \tag{4.14}$$

with

$$A(t) = \begin{bmatrix} 0 & 0 & 0 & 0 & 0 & 0 \\ 0 & 0 & 0 & 0 & 0 & 0 \\ 0 & u_{d1}(t) & 0 & 0 & 0 & 0 \\ 0 & 0 & 0 & 0 & 0 & 0 \\ 0 & 0 & 0 & u_{d1}(t) & 0 & 0 \\ 0 & 0 & 0 & 0 & 0 & 0 \end{bmatrix}$$

and

$$B(t) = \begin{bmatrix} 1 & 0 & 0 & 0 \\ 0 & 1 & 0 & 0 \\ x_{d2}(t) & 0 & 0 & 0 \\ 0 & 0 & 1 & 0 \\ x_{d4}(t) & 0 & 0 & 0 \\ 0 & 0 & 0 & 1 \end{bmatrix}$$

The system given by Equation 4.14 is linear time varying and can easily be proven to be controllable by checking its Grammian [21] to be nonsingular. For a linear trajectory with constant velocity $u_{d1}(t) = u_{d1}$ the controllability condition is given by

$$rank\{B, AB, A^2B, A^3B, A^4B, A^5B\} = 6 \tag{4.15}$$

The matrix in Equation 4.15 is a nonsingular matrix and has at least one nonzero minor of order six. The controllability matrix is nonsingular only as long as the input u_{d1} to the system is nonzero. This corresponds to the singularity in the kinematic model of the system. Thus, the system is controllable as long as $u_{d1} \neq 0$. Choosing the linear time-varying feedback law for the system as

$$u_e = -Kx_e \tag{4.16}$$

for the chained-form system the control law should be such that the feedback law for each chain contains the same number of terms as the number of states in that chain. Thus,

$$\begin{aligned} u_{e1} &= -k_1 x_{e1} \\ u_{e2} &= -k_2 x_{e2} - \frac{k_3}{u_{d1}} x_{e3} \\ u_{e3} &= -k_4 x_{e4} - \frac{k_5}{u_{d1}} x_{e5} \\ u_{e4} &= -k_6 x_{e6} \end{aligned} \tag{4.17}$$

The feedback coefficients k_3 and k_5 are divided by u_{d1} so that the characteristic equation of the closed-loop system matrix does not contain u_{d1}, thus making the design global. Thus, the matrix K is given as

$$u_{e4} = \begin{bmatrix} k_1 & 0 & 0 & 0 & 0 & 0 \\ 0 & k_2 & k_3/u_{d1} & 0 & 0 & 0 \\ 0 & 0 & 0 & k_4 & k_5/u_{d1} & 0 \\ 0 & 0 & 0 & 0 & 0 & k_6 \end{bmatrix} \tag{4.18}$$

The k's are chosen such that k_1 and k_6 are positive, and k_2, k_3, k_4, and k_5 are such that $\lambda^2 + k_2\lambda + k_3$ and $\lambda^2 + k_4\lambda + k_5$ are Hurwitz. The closed-loop system matrix is thus given as

$$A_{cl} = A - BK$$

$$= \begin{bmatrix} k_1 & 0 & 0 & 0 & 0 & 0 \\ 0 & -k_2 & -k_3/u_{d1} & 0 & 0 & 0 \\ x_{d2}k_1 & u_{d1} & 0 & 0 & 0 & 0 \\ 0 & 0 & 0 & -k_4 & -k_5/u_{d1} & 0 \\ x_{d4}k_1 & 0 & 0 & u_{d1} & 0 & 0 \\ 0 & 0 & 0 & 0 & 0 & k_6 \end{bmatrix} \tag{4.19}$$

The closed-loop system matrix has constant eigenvalues with negative real parts. This does not guarantee the asymptotic stability of the closed-loop time-varying system [26]. However, for specific choices of $u_{d1}(t)$, bounded away from zero and $u_{d2}(t)$, $u_{d3}(t)$, and $u_{d4}(t) = 1$, it is possible to use results on slowly varying linear systems in order to prove asymptotic stability. The feedback matrix is obtained by using the pole placement technique. Since for stability the eigenvalues of the closed-loop matrix should have negative real parts, the characteristic equation of the system should satisfy the following:

$$\det(sI - A_{cl}) = (s - p_1)(s - p_2)(s - p_3)(s - p_4)(s - p_5)(s - p_6)$$

where p_i, $i = 1, 2, \ldots, 6$ are the eigenvalues of the system. The resulting closed-loop system (Equation 4.19) is controllable with the choice of feedback in (Equation 4.17).

4.3.1 Simulation of the Controller

For simulation, the following sinusoidal trajectory is chosen:

$$x_d(t) = t \quad y_d(t) = a \sin \omega t \quad z_d(t) = 1$$

which gives the desired values for the chained-form states and inputs as follows:

$$x_{d1}(t) = t \quad x_{d2}(t) = a\omega \cos \omega t \quad x_{d3}(t) = a \sin \omega t$$

$$x_{d4}(t) = 0 \quad x_{d5}(t) = 1 \quad x_{d6}(t) = 0$$

and

$$u_{d1}(t) = 1 \quad u_{d2}(t) = -a\omega^2 \sin \omega t$$

$$u_{d3}(t) = 0 \quad u_{d4}(t) = 0$$

The initial conditions for the states are

$$x_{d1}(0) = 1 \quad x_{d2}(0) = a\omega$$

$$x_{d3}(0) = 2 \quad x_{d4}(0) = 0$$

$$x_{d5}(0) = 3 \quad x_{d6}(0) = 0$$

Choosing the six coincident closed-loop poles at –2, that is,

$$p_1 = p_2 = p_3 = p_4 = p_5 = p_6 = -2$$

we get the feedback matrix coefficients as

$$k_1 = k_6 = 2 \quad \text{and} \quad k_2 = k_3 = k_4 = k_5 = 4$$

Choosing $a = 1$ and $\omega = \pi$, we get the simulation results. The results show the tracking errors for chained-form states and inputs, and for actual states and inputs. Once the tracking errors go to zero, the actual control inputs as obtained from the chained-form variables and inputs are the same as the computed desired inputs. The desired inputs are computed from actual system variables from Equation 4.5. Since the control design is based on the linearization of the system, the controller will make the controlled system locally asymptotically stable. The simulation results are presented in Figure 4.1 through 4.12.

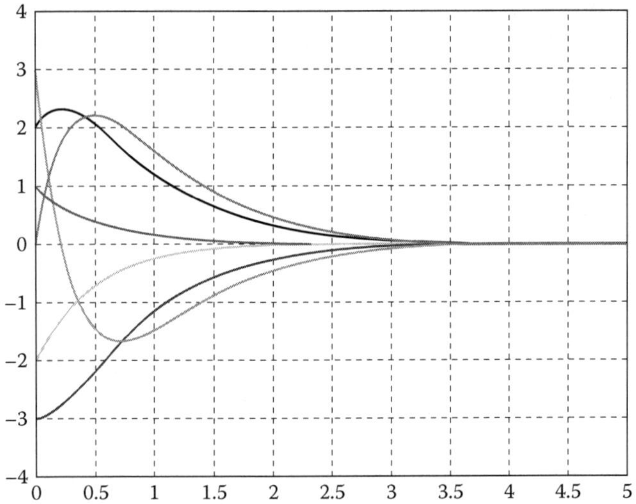

FIGURE 4.1 The result of approximate linearization: tracking errors in chained-form variables vs. time (s).

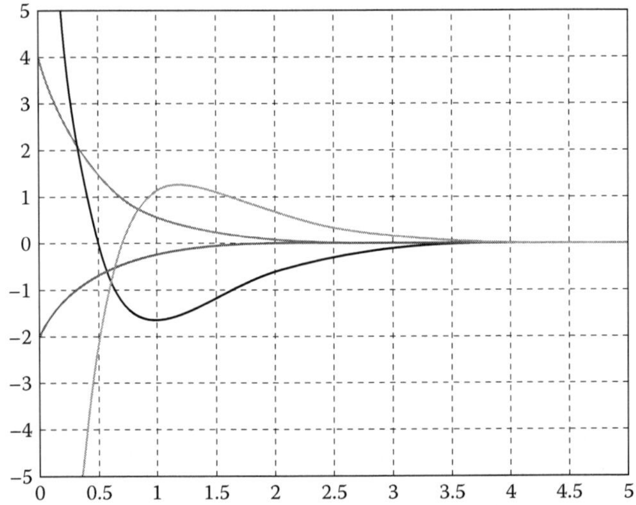

FIGURE 4.2 The result of approximate linearization: tracking errors (m/s) in chained-form inputs vs. time (s).

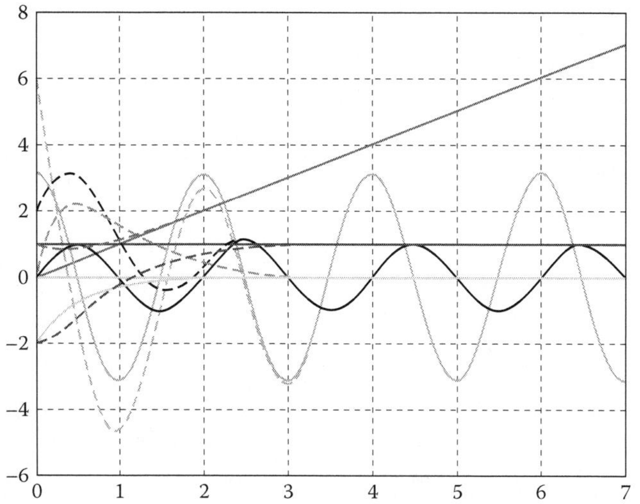

FIGURE 4.3 The result of approximate linearization: actual (--) and desired (-) chained-form variables vs. time (s).

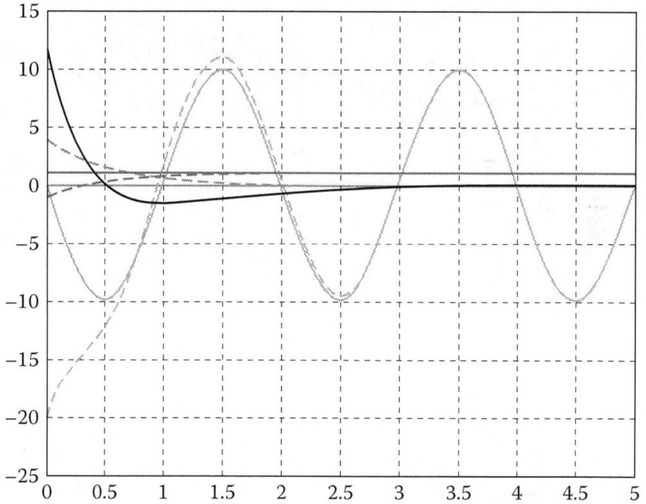

FIGURE 4.4 The result of approximate linearization: actual (--) and desired (-) chained-form inputs vs. time (s).

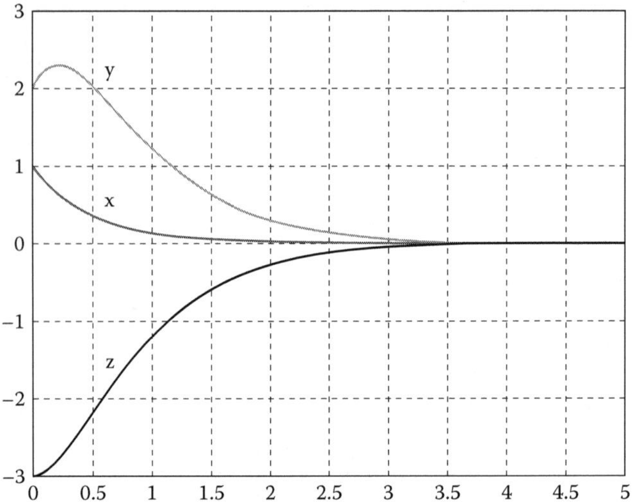

FIGURE 4.5 The result of approximate linearization: tracking errors (*m*) in variables *x*, *y*, and *z* vs. time (*s*).

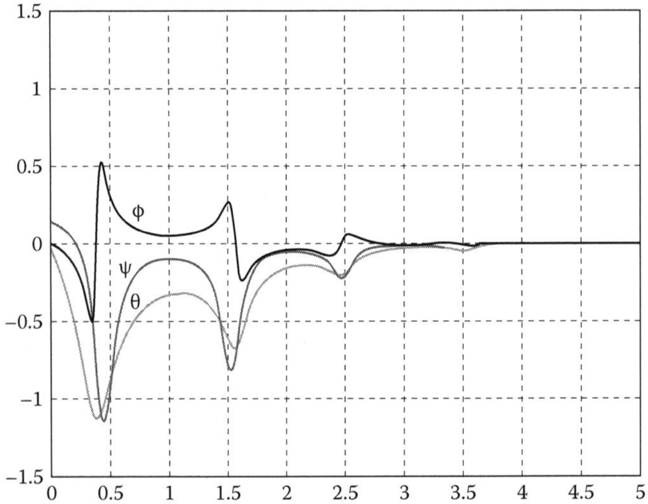

FIGURE 4.6 The result of approximate linearization: tracking errors (rad) in variables ψ, θ, and φ vs. time (*s*).

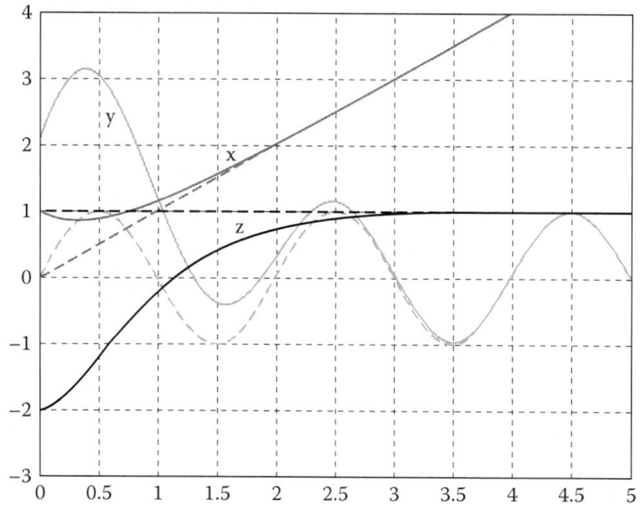

FIGURE 4.7 The result of approximate linearization: actual (--) and desired (-) original variables x, y, and z vs. time (s).

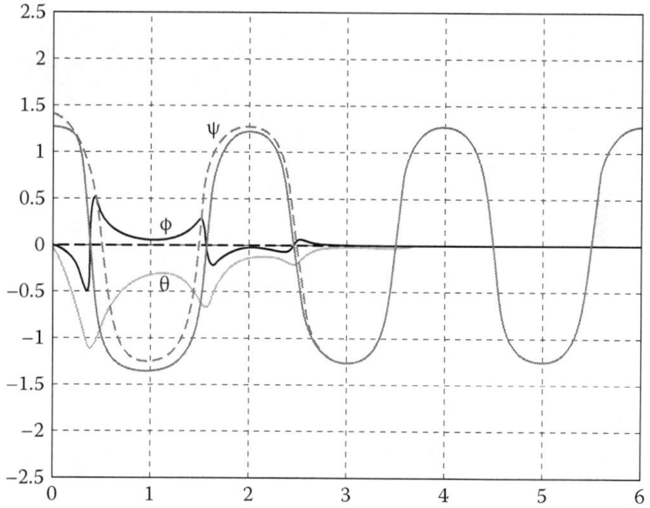

FIGURE 4.8 The result of approximate linearization: actual (--) and desired (-) original variables ψ, θ, and φ, vs. time (s).

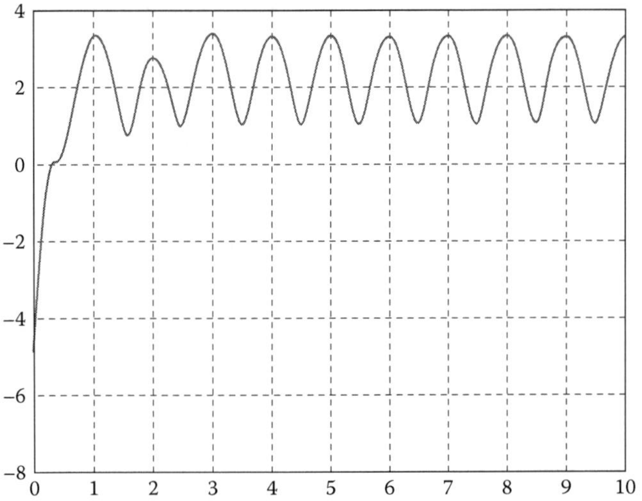

FIGURE 4.9 The result of approximate linearization: v_1 (m/s) vs. time (s).

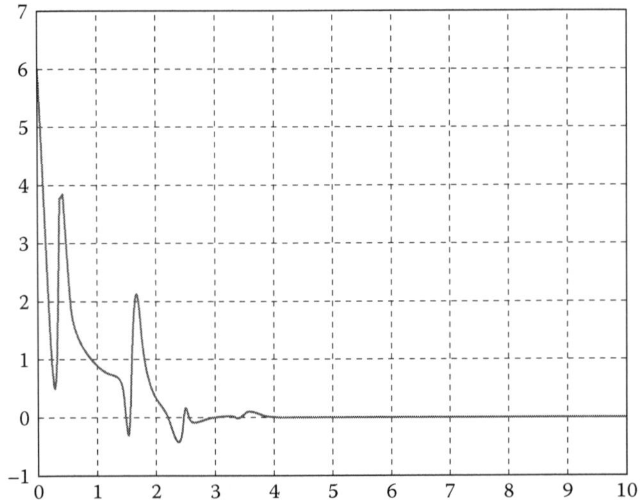

FIGURE 4.10 The result of approximate linearization: v_2 (rad/s) vs. time (s).

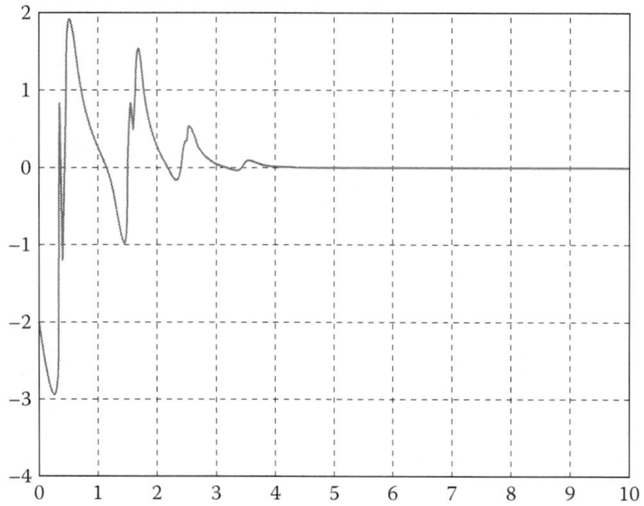

FIGURE 4.11 The result of approximate linearization: v_3 (rad/s) vs. time (s).

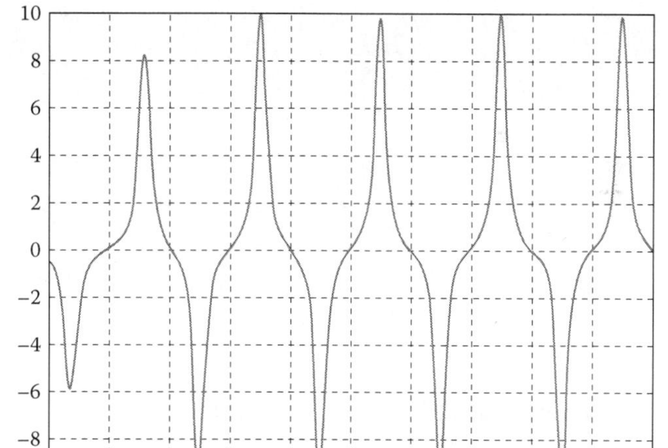

FIGURE 4.12 The result of approximate linearization: v_4 (rad/s) vs. time (s).

4.3.2 MATLAB® Program Code for the Approximate Linearization

```
%%%%%%%%%%%%%%%%%%%%%%%%%%%%%%%%
% Approximate Linearization for kinematic model, Main
code %
%%%%%%%%%%%%%%%%%%%%%%%%%%%%%%%%%
t=[0:0.01:10]';a=1;w=pi; xd=t; xddot=1;
yddot=a*w*cos(w*t); y2ddot=-a*w*w*sin(w*t); yd=a*sin(w*t);
den=(sqrt(xddot.*xddot+yddot.*yddot)).*(sqrt(xddot.*xddo
t+yddot.*yddot)).*(sqrt(xddot.*xddot+yddot.*yddot));
 [t,x]=ode45('F',[t],[-2,1,0]);
 [t,y]=ode45('F1',[t],[-1,a*w,0]);
xe=-x(:,1)+xd; ye=-y(:,1)+yd;
theta=atan(y(:,2)./x(:,2)); thetad=atan(a*w*cos(w*t));th
etae=thetad-theta;
x2=(x(:,2).*y(:,3)-x(:,3).*y(:,2))./
(x(:,1).*x(:,1).*x(:,1));
phi=atan(cos(theta).*cos(theta).*cos(theta).*x2);
phid=atan((y2ddot.*xddot)./den); phie=phid-phi;
%%%%%%%%%%%%%%%%%%%%%%%%%%
% Matlab Functions 'F' and 'F1'%
%%%%%%%%%%%%%%%%%%%%%%%%%%
function xp = F(t,x)
a=1; w=pi;
xp = zeros(3,1);
xp(1)=x(2);
xp(2)=x(3);
xp(3)=-15*x(3)-75*x(2)-125*x(1)+75+125*t;
function yp = F1(t,y)
a=1; w=pi;
yp = zeros(3,1);
yp(1)=y(2);
yp(2)=y(3);
yp(3)=-15*y(3)-75*y(2)-125*y(1)+75*a*w*cos(w*t)-
15*a*w*w*sin(w*t)- a*w*w*w*cos(w*t)+125*a*sin(w*t);
%%%%%%%%%%%%%%%%%%%%%%%%%%%%%%%%%%%%%%%%
% Computation of tracking errors in chained form
variables %
%%%%%%%%%%%%%%%%%%%%%%%%%%%%%%%%%%%%%%%%
%%% Original Syatem %%%
A=[0 0 0 0 0 0;0 0 0 0 0 0;0 1 0 0 0 0; 0 0 0 0 0 0 ; 0
0 0 1 0 0;0 0 0 0 0 0];
B=[1 0 0 0;0 1 0 0 ;0 0 0 0;0 0 1 0; 0 0 0 0; 0 0 0 1];
% %% Pole locations %%%
s1=s2=s3=s4=s5=s6=-2; wn=2; qsi=1;
```

```
%%% Feedback system %%%%
K=[2 0 0 0 0 0; 0 4 4 0 0 0; 0 0 0 4 4 0; 0 0 0 0 0 2];
Acl = A - B*K; C=zeros(6,6); D=zeros(6,4);
sys=ss(Acl,B,C,D);
% %% response for the chained form error variables %%%
pi=22/7; a=1; w=pi; t=[0:0.01:9.999];
xe0=[0;a*w;0;0;1;0];    %%%on trajectory initially
xe0=[1;a*w;2;0;-3;-2]; %% %off trajectory initially
xef=[0;0;0;0;0;0];
ue1=-K(1,1)*xef(1); ue2=-K(2,2)*xef(2)-K(2,3)*xef(3);
ue3=-K(3,4)*xef(4)-K(3,5)*xef(5); ue4=-K(4,6)*xef(6);
ue=[ue1;ue2;ue3;ue4]; ue=ue*ones(1,1000);
[y,t,x]=lsim(sys,ue,t,xe0);
Figure (1);
plot(t,x(:,[1,3,5])),grid, title('Error in chained form
states 1,3,and 5:xe1(t),xe3(t),xe5(t)');
hold on;
plot(t,x(:,[2,4,6])),grid, title('Error in chained form
states 2,4,and 6:xe2(t),xe4(t),xe6(t)');
hold off;
%%%%%%%%%%%%%%%%%%%%%%%%%%%%%%%%%%%%%
% Computation of tracking errors in chained form inputs %
%%%%%%%%%%%%%%%%%%%%%%%%%%%%%%%%%%%%%
A=[0 0 0 0 0 0;0 0 0 0 0 0;0 1 0 0 0 0; 0 0 0 0 0 0 ; 0
0 0 1 0 0;0 0 0 0 0 0];
B=[1 0 0 0;0 1 0 0 ;0 0 0 0;0 0 1 0; 0 0 0 0; 0 0 0 1];
% %% Pole locations %%%
s1=s2=s3=s4=s5=s6=-2; wn=2; qsi=1;
%%% Feedback system %%%%
K=[2 0 0 0 0 0; 0 4 4 0 0 0; 0 0 0 4 4 0; 0 0 0 0 0 2];
Acl = A - B*K; C=zeros(6,6); D=zeros(6,4);
sys=ss(Acl,B,C,D);
pi=22/7; a=1; w=pi; t=[0:0.01:9.999];
xe0=[0;a*w;0;0;1;0];     %%%on trajectory initially
xe0=[1;a*w;2;0;-3;-2];  %% %off trajectory initially
xef=[0;0;0;0;0;0];
ue1=-K(1,1)*xef(1); ue2=-K(2,2)*xef(2)-K(2,3)*xef(3);
ue3=-K(3,4)*xef(4)-K(3,5)*xef(5); ue4=-K(4,6)*xef(6);
ue=[ue1;ue2;ue3;ue4]; ue=ue*ones(1,1000);
[y,t,x]=lsim(sys,ue,t,xe0); xa=x';
% %% response for the error in chained form inputs %%%
ur1=-K(1,1)*xa(1,:);
ur2=-K(2,2)*xa(2,:)-K(2,3)*xa(3,:);
ur3=-K(3,4)*xa(4,:)-K(3,5)*xa(5,:);
ur4=-K(4,6)*xa(6,:); ur=[ur1;ur2;ur3;ur4];
```

```
Figure (2);
Plot (t,ur), grid, title('Error in inputs for chained
form vs time ');
plot(t,ur(1,:),'g'),grid; hold on;
plot(t,ur(2,:),'c'),
plot(t,ur(3,:),'b'); plot(t,ur(4,:),'r'); hold off
%%%%%%%%%%%%%%%%%%%%%%%
% Actual and desired chained form variables %
%%%%%%%%%%%%%%%%%%%%%%%%
%%% Feedback system %%%%
K=[2 0 0 0 0 0; 0 4 4 0 0 0; 0 0 0 4 4 0; 0 0 0 0 0 2];
Acl = A - B*K; C=zeros(6,6); D=zeros(6,4);
sys=ss(Acl,B,C,D);
pi=22/7; a=1; w=pi; t=[0:0.01:9.999];
xe0=[0;a*w;0;0;1;0];    %%%on trajectory initially
xe0=[1;a*w;2;0;-3;-2];  %% %off trajectory initially
xef=[0;0;0;0;0;0];
ue1=-K(1,1)*xef(1); ue2=-K(2,2)*xef(2)-K(2,3)*xef(3);
ue3=-K(3,4)*xef(4)-K(3,5)*xef(5); ue4=-K(4,6)*xef(6);
ue=[ue1;ue2;ue3;ue4]; ue=ue*ones(1,1000);
[y,t,x]=lsim(sys,ue,t,xe0);
%%% response for the chained form actual and desired
variables %%%
xd1=t; xd2=a*w*cos(w*t); xd3=a*sin(w*t);
xd4=0*ones(1,1000); xd5=1*ones(1,1000);
xd6=0*ones(1,1000);
xd=[xd1';xd2';xd3';xd4;xd5;xd6];
xact=x'+xd;
Figure (3);
plot(t,xd); grid,title('Chained form desired states vs
time'); hold on;
plot(t,xact,'--'); grid,title('Chained form actual
states vs time,'); hold off;
%%%%%%%%%%%%%%%%%%%%%%%%%%%%%%
% Computation of actual and desired chained form inputs %
%%%%%%%%%%%%%%%%%%%%%%%%%%%%%%
A=[0 0 0 0 0 0;0 0 0 0 0 0;0 1 0 0 0 0; 0 0 0 0 0 0 ; 0
0 0 1 0 0;0 0 0 0 0 0];
B=[1 0 0 0;0 1 0 0 ;0 0 0 0;0 0 1 0; 0 0 0 0; 0 0 0 1];
% %% Pole locations %%%
s1=s2=s3=s4=s5=s6=-2; wn=2; qsi=1;
%%% Feedback system %%%%
K=[2 0 0 0 0 0; 0 4 4 0 0 0; 0 0 0 4 4 0; 0 0 0 0 0 2];
Acl = A - B*K; C=zeros(6,6); D=zeros(6,4);
sys=ss(Acl,B,C,D);
pi=22/7; a=1; w=pi; t=[0:0.01:9.999];
```

```
xe0=[0;a*w;0;0;1;0];     %%%on trajectory initially
xe0=[1;a*w;2;0;-3;-2];   %% %off trajectory initially
xef=[0;0;0;0;0;0];
ue1=-K(1,1)*xef(1); ue2=-K(2,2)*xef(2)-K(2,3)*xef(3);
ue3=-K(3,4)*xef(4)-K(3,5)*xef(5); ue4=-K(4,6)*xef(6);
ue=[ue1;ue2;ue3;ue4]; ue=ue*ones(1,1000);
[y,t,x]=lsim(sys,ue,t,xe0); xa=x';
%%% response for the chained form actual and desired
inputs%%%
ur1=-K(1,1)*xa(1,:);
ur2=-K(2,2)*xa(2,:)-K(2,3)*xa(3,:);
ur3=-K(3,4)*xa(4,:)-K(3,5)*xa(5,:);
ur4=-K(4,6)*xa(6,:);
ur=[ur1;ur2;ur3;ur4];
ud1=1; ud1n=ud1*ones(1,1000);
ud2=-a*w^2*sin(w*t); ud2n=ud2';
ud3=0*ones(1,1000); ud4=0*ones(1,1000);
ud=[ud1n;ud2n;ud3;ud4];
u=ud+ur;
Figure (4);
plot(t,ud), grid, title('Chained form desired and actual
inputs vs time,'); hold on; grid on
plot(t,u,'--'); hold off
%%%%%%%%%%%%%%%%%%%%%%%%%%%%
% Computation of tracking errors in variables x ,y , z %
%%%%%%%%%%%%%%%%%%%%%%%%%%%%%
%Feedback system
K=[2 0 0 0 0 0; 0 4 4 0 0 0; 0 0 0 4 4 0; 0 0 0 0 0
2];%roots are (-2)
Acl = A - B*K; C=zeros(6,6); D=zeros(6,4);
sys=ss(Acl,B,C,D);
pi=22/7; a=1; w=pi;t=[0:0.01:9.999];
xe0=[0;a*w;0;0;1;0];%on trajectory initially
xe0=[1;a*w;2;0;-3;-2];%off trajectory initially
xef=[0;0;0;0;0;0];
ue1=-K(1,1)*xef(1);
ue2=-K(2,2)*xef(2)-K(2,3)*xef(3);
ue3=-K(3,4)*xef(4)-K(3,5)*xef(5);
ue4=-K(4,6)*xef(6);
ue=[ue1;ue2;ue3;ue4]; ue=ue*ones(1,1000);
[y,t,x]=lsim(sys,ue,t,xe0);
xd1=t; xd2=a*w*cos(w*t); xd3=a*sin(w*t);
xd4=0*ones(1,1000); xd5=1*ones(1,1000);
xd6=0*ones(1,1000);
xd=[xd1';xd2';xd3';xd4;xd5;xd6];
xact=x'+xd;
```

```
%%%response for the errors in actual system variables%%%
X=xact(1,:);
Y=xact(3,:);
Z=xact(5,:);
W1=[X;Y;Z];
Xd=xd(1,:); Yd=xd(3,:); Zd=xd(5,:);
Wd1=[Xd;Yd;Zd];
We1=W1-Wd1;
Figure (5);
plot(t,We1), grid, title('Error in original state
variables, x,y,z '); grid on
%%%%%%%%%%%%%%%%%%%%%%%%%%%%%%%%%
% Computation of tracking errors in variables ψ, θ, φ %
%%%%%%%%%%%%%%%%%%%%%%%%%%%%%%%%%
%Feedback system
K=[2 0 0 0 0 0; 0 4 4 0 0 0; 0 0 0 4 4 0; 0 0 0 0 0 2];
Acl = A - B*K; C=zeros(6,6); D=zeros(6,4);
sys=ss(Acl,B,C,D);
pi=22/7; a=1; w=pi;t=[0:0.01:9.999];
xe0=[0;a*w;0;0;1;0];%on trajectory initially
xe0=[1;a*w;2;0;-3;-2];%off trajectory initially
xef=[0;0;0;0;0;0];
ue1=-K(1,1)*xef(1);
ue2=-K(2,2)*xef(2)-K(2,3)*xef(3);
ue3=-K(3,4)*xef(4)-K(3,5)*xef(5);
ue4=-K(4,6)*xef(6);
ue=[ue1;ue2;ue3;ue4]; ue=ue*ones(1,1000);
[y,t,x]=lsim(sys,ue,t,xe0);
xd1=t; xd2=a*w*cos(w*t); xd3=a*sin(w*t);
xd4=0*ones(1,1000); xd5=1*ones(1,1000);
xd6=0*ones(1,1000);
xd=[xd1';xd2';xd3';xd4;xd5;xd6];
xact=x'+xd;
%%% response for the errors in actual system variables
%%%
X=xact(1,:);
Y=xact(3,:);
Z=xact(5,:);
W1=[X;Y;Z];
PSI=atan(xact(2,:));
theta=-atan(xact(4,:)./sec(PSI));
phi=atan (xact(6,:));
phi=acot(cot(theta)./sin(PSI)+tan(PSI)./sin(theta));
W2=[PSI;theta;phi];
PSId=atan(xd(2,:));
```

```
thetad=-atan(xd(4,:)./sec(PSId));
phid=atan (xd(6,:));
phid=acot(cot(thetad)./sin(PSId)+tan(PSId)./sin(thetad)) ;
Wd2=[PSId;thetad;phid];
We2=W2-Wd2;
Figure (6);
plot(t,We2),grid; title('Error in original state
variables,(psie,thetae,phie)');
%%%%%%%%%%%%%%%%%%%%%%%%%%%%%%%%
% Actual and desired original variables x, y, z %
%%%%%%%%%%%%%%%%%%%%%%%%%%%%%%%%
K=[2 0 0 0 0 0; 0 4 4 0 0 0; 0 0 0 4 4 0; 0 0 0 0 0 2];
Acl = A - B*K; C=zeros(6,6); D=zeros(6,4);
sys=ss(Acl,B,C,D);
pi=22/7; a=1; w=pi;t=[0:0.01:9.999];
xe0=[0;a*w;0;0;1;0];%on trajectory initially
xe0=[1;a*w;2;0;-3;-2];%off trajectory initially
xef=[0;0;0;0;0;0];
ue1=-K(1,1)*xef(1);
ue2=-K(2,2)*xef(2)-K(2,3)*xef(3);
ue3=-K(3,4)*xef(4)-K(3,5)*xef(5);
ue4=-K(4,6)*xef(6);
ue=[ue1;ue2;ue3;ue4]; ue=ue*ones(1,1000);
[y,t,x]=lsim(sys,ue,t,xe0);
xd1=t;
xd2=a*w*cos(w*t);
xd3=a*sin(w*t);
xd4=0*ones(1,1000);
xd5=1*ones(1,1000);
xd6=0*ones(1,1000);
xd=[xd1';xd2';xd3';xd4;xd5;xd6];
xact=x'+xd;
X=xact(1,:);
Y=xact(3,:);
Z=xact(5,:);
W1=[X;Y;Z];
Xd=xd(1,:);
Yd=xd(3,:);
Zd=xd(5,:);
Wd1=[Xd;Yd;Zd];
Figure (7);
plot(t,W1),grid,title('Original system actual state
variables,(x,y,z)');hold on;
plot(t,Wd1,'--'); title('Original system actual state
variables,(xd,yd,zd)');hold off
```

```
%%%%%%%%%%%%%%%%%%%%%%%%%
% Actual and desired original variables ψ, θ, φ %
%%%%%%%%%%%%%%%%%%%%%%%%%
K=[2 0 0 0 0 0; 0 4 4 0 0 0; 0 0 0 4 4 0; 0 0 0 0 0 2];
Acl = A - B*K; C=zeros(6,6); D=zeros(6,4);
sys=ss(Acl,B,C,D);
pi=22/7; a=1; w=pi;t=[0:0.01:9.999];
xe0=[0;a*w;0;0;1;0];%on trajectory initially
xe0=[1;a*w;2;0;-3;-2];%off trajectory initially
xef=[0;0;0;0;0;0];
ue1=-K(1,1)*xef(1);
ue2=-K(2,2)*xef(2)-K(2,3)*xef(3);
ue3=-K(3,4)*xef(4)-K(3,5)*xef(5);
ue4=-K(4,6)*xef(6);
ue=[ue1;ue2;ue3;ue4]; ue=ue*ones(1,1000);
[y,t,x]=lsim(sys,ue,t,xe0);
xd1=t;
xd2=a*w*cos(w*t);
xd3=a*sin(w*t);
xd4=0*ones(1,1000);
xd5=1*ones(1,1000);
xd6=0*ones(1,1000);
xd=[xd1';xd2';xd3';xd4;xd5;xd6];
xact=x'+xd;
X=xact(1,:);
Y=xact(3,:);
Z=xact(5,:);
W1=[X;Y;Z];
Xd=xd(1,:);
Yd=xd(3,:);
Zd=xd(5,:);
Wd1=[Xd;Yd;Zd];
PSI=atan(xact(2,:));
theta=-atan(xact(4,:)./sec(PSI));
phi=atan (xact(6,:));
phi=acot(cot(theta)./sin(PSI)+tan(PSI)./sin(theta));
W2=[PSI;theta;phi];
PSId=atan(xd(2,:));
thetad=-atan(xd(4,:)./sec(PSId));
phid=atan (xd(6,:));
phid=acot(cot(thetad)./sin(PSId)+tan(PSId)./sin(thetad)) ;
Wd2=[PSId;thetad;phid];
We2=W2-Wd2;
Figure (8);
plot(t,W2),grid,title('Original system actual state
variables, (psi,theta,phi) ');hold on;
```

```
plot(t,Wd2,'--'); title('Original system actual state va
riables,((psid,thetad,phid)');hold off;

%%%%%%%%%%%%%%%%%%%%%%
% Computation of actual input v1%
%%%%%%%%%%%%%%%%%%%%%%

A=[0 0 0 0 0 0;0 0 0 0 0 0;0 1 0 0 0 0; 0 0 0 0 0 0 ; 0
0 0 1 0 0;0 0 0 0 0 0];
B=[1 0 0 0;0 1 0 0 ;0 0 0 0;0 0 1 0; 0 0 0 0; 0 0 0 1];
% %% Pole locations %%%
s1=s2=s3=s4=s5=s6=-2; wn=2; qsi=1;
%%% Feedback system %%%%
K=[2 0 0 0 0 0; 0 4 4 0 0 0; 0 0 0 4 4 0; 0 0 0 0 0 2];
Acl = A - B*K; C=zeros(6,6); D=zeros(6,4);
sys=ss(Acl,B,C,D);
pi=22/7; a=1; w=pi; t=[0:0.01:9.999];
xe0=[0;a*w;0;0;1;0];   %%%on trajectory initially
xe0=[1;a*w;2;0;-3;-2]; %% %off trajectory initially
xef=[0;0;0;0;0;0];
ue1=-K(1,1)*xef(1); ue2=-K(2,2)*xef(2)-K(2,3)*xef(3);
ue3=-K(3,4)*xef(4)-K(3,5)*xef(5); ue4=-K(4,6)*xef(6);
ue=[ue1;ue2;ue3;ue4]; ue=ue*ones(1,1000);
[y,t,x]=lsim(sys,ue,t,xe0); xa=x';
xd1=t;
xd2=a*w*cos(w*t);
xd3=a*sin(w*t);
xd4=0*ones(1,1000);
xd5=1*ones(1,1000);
xd6=0*ones(1,1000);
xd=[xd1';xd2';xd3';xd4;xd5;xd6];
xact=x'+xd;
ur1=-K(1,1)*xa(1,:);
ur2=-K(2,2)*xa(2,:)-K(2,3)*xa(3,:);
ur3=-K(3,4)*xa(4,:)-K(3,5)*xa(5,:);
ur4=-K(4,6)*xa(6,:);
ur=[ur1;ur2;ur3;ur4];
ud1=1;
ud1n=ud1*ones(1,1000);
ud2=-a*w^2*sin(w*t);
ud2n=ud2';
ud3=0*ones(1,1000);
ud4=0*ones(1,1000);
ud=[ud1n;ud2n;ud3;ud4];
u=ud+ur;
PSI=atan(xact(2,:));
```

```
theta=-atan(xact(4,:)./sec(PSI));
phi=atan (xact(6,:));
phi=acot(cot(theta)./sin(PSI)+tan(PSI)./sin(theta));
v1=u(1,:)./cos(PSI).*cos(theta);
Figure (9);
plot(t,v1),grid, title( 'original system input1 (v1');
%%%%%%%%%%%%%%%%%%%%%%%%%
% Computation of input v2 %
%%%%%%%%%%%%%%%%%%%%%%%%%%
r11=cos(PSI).*cos(theta);
r23=sin(PSI).*sin(theta).*cos(phi)-cos(PSI).*sin(phi);
r33=cos(theta).*cos(phi);
r22=sin(PSI).*sin(theta).*sin(phi)+cos(PSI).*cos(phi);
r32=sin(phi).*cos(theta);
r12=cos(PSI).*sin(theta).*sin(phi)-sin(PSI).*cos(phi);
r13=sin(PSI).*sin(theta).*cos(phi)+sin(PSI).*sin(phi);
R=r11+r22+r33;
R1=1+R;
r11n=1./(1+r11);
v3=r11.*(-r23.*u(2,:)-r33.*u(3,:));
v4=r11.*(r22.*u(2,:)+r32.*u(3,:));
v2=r11n.*(R1.*u(4,:)-r12.*v3-r13.*v4);
Figure (10);
plot(t,v2),grid, title( 'original system input2 (v2');
%%%%%%%%%%%%%%%%%%%%%%%%%
% Computation of input v3 %
%%%%%%%%%%%%%%%%%%%%%%%%%
r11=cos(PSI).*cos(theta);
r23=sin(PSI).*sin(theta).*cos(phi)-cos(PSI).*sin(phi);
r33=cos(theta).*cos(phi);
r22=sin(PSI).*sin(theta).*sin(phi)+cos(PSI).*cos(phi);
r32=sin(phi).*cos(theta);
r12=cos(PSI).*sin(theta).*sin(phi)-sin(PSI).*cos(phi);
r13=sin(PSI).*sin(theta).*cos(phi)+sin(PSI).*sin(phi);
R=r11+r22+r33;
R1=1+R;
r11n=1./(1+r11);
v3=r11.*(-r23.*u(2,:)-r33.*u(3,:));
Figure (11);
plot(t,v3),grid, title( 'original system input3 (v3');
%%%%%%%%%%%%%%%%%%%%%%%%%
% Computation of input v4 %
%%%%%%%%%%%%%%%%%%%%%%%%%%
r11=cos(PSI).*cos(theta);
```

```
r23=sin(PSI).*sin(theta).*cos(phi)-cos(PSI).*sin(phi);
r33=cos(theta).*cos(phi);
r22=sin(PSI).*sin(theta).*sin(phi)+cos(PSI).*cos(phi);
r32=sin(phi).*cos(theta);
r12=cos(PSI).*sin(theta).*sin(phi)-sin(PSI).*cos(phi);
r13=sin(PSI).*sin(theta).*cos(phi)+sin(PSI).*sin(phi);
R=r11+r22+r33;
R1=1+R;
r11n=1./(1+r11);
v4=r11.*(r22.*u(2,:)+r32.*u(3,:));
Figure (12);
plot(t,v4),grid, title( 'original system input4 (v4)');
```

4.4 CONTROL USING EXACT FEEDBACK LINEARIZATION VIA STATE AND INPUT TRANSFORMATIONS

In this section the use of nonlinear feedback design is used for the global stabilization of the tracking error associated with the trajectory to zero. For nonlinear systems two types of exact linearization methods are generally used. One is the full-state feedback transformation of the differential equations of the system into the linear system. Another is the input–output linearization, which results in the input–output differential map being linear. Both the feedback problems can be solved using either the static or the dynamic feedback.

For the nonholonomic driftless system $\dot{q}=G(q)v$ the full-state linearization of the system cannot be achieved by using a smooth static (time-invariant) state feedback. The reason for this is in fact the controllability condition given by $rank[g_1,g_2,g_3,g_4,[g_1,g_3][g_1,g_4]]=6$, which means that the distribution generated by the vector fields g_1,g_2,g_3,g_4 is not involutive, which violates the necessary condition for full-static-state feedback linearization [13]. Thus, for exact linearization of the system, the method of dynamic state feedback is used.

For the above nonlinear system, dynamic feedback linearization consists of finding a dynamic feedback compensator of the form

$$\begin{aligned}\dot{\zeta}&=a(q,\zeta)+b(q,\zeta)r\\ u&=c(q,\zeta)+d(q,\zeta)r\end{aligned}\tag{4.20}$$

The state vector ζ is the compensator state whose dimensions depend upon the number of integrators added on the input channels. The vector r is the auxiliary input vector, which is the new input to the integrators added.

The starting point of the dynamic extension for our problem is to define an m dimensional output vector $z=h(q)$, m being 4 in this case. A certain desired behavior is assigned to this output vector. The output vector is then successively differentiated until each and every input in the system appears and the invertible map (or matrix) is nonsingular. During the successive differentiations of the output vector it becomes necessary to add the chain of integrators on the inputs so as to avoid their direct differentiation. The number of integrators results in the compensator state vector ζ. The input to these integrators becomes the new auxiliary input r. The process continues and terminates after a finite number of differentiations if the system is invertible from the chosen output vector z. If the sum of the orders of the output differentiations is equal to the sum of the order of the original system (n) and the dimensions of the compensator, then the full-state linearization is achieved in the sense that no internal dynamics are left in the system. The process also results in the decoupling of the output vector from the new auxiliary input.

If at some point of differentiation of output in the algorithm the decoupling matrix of the system is nonsingular without the addition of any compensator state, the process results in the input–output linearization of the system. The static feedback law of the form

$$u = a(q) + b(q)r \tag{4.21}$$

is used to linearize the system.

4.4.1 Control Using Exact Feedback Linearization via Static Feedback

For our system let the output vector in the chained form be defined as

$$z = \begin{pmatrix} x_1 \\ x_3 \\ x_5 \\ x_6 \end{pmatrix} \tag{4.22}$$

The derivative of the output is given as

$$\dot{z} = \begin{pmatrix} \dot{x}_1 \\ \dot{x}_3 \\ \dot{x}_5 \\ \dot{x}_6 \end{pmatrix} = \begin{pmatrix} 1 & 0 & 0 & 0 \\ x_2 & 0 & 0 & 0 \\ x_4 & 0 & 0 & 0 \\ 0 & 0 & 0 & 1 \end{pmatrix} \begin{pmatrix} u_1 \\ u_2 \\ u_3 \\ u_4 \end{pmatrix}$$

or

$$\dot{z} = H(q)u \tag{4.23}$$

The inputs u_2 and u_3 do not appear after the differentiation, and also the decoupling matrix $H(q)$ is singular and not invertible. Hence, the static feedback cannot be applied and the system cannot be linearized by input–output linearization. The same result follows if the above procedure is repeated by choosing the actual state variable as the output vector.

4.4.2 Control Using Exact Feedback Linearization via Dynamic Feedback

Since the static feedback fails to solve the problem, we will be making use of dynamic feedback extension. For the linearization via dynamic feedback let us again define the linearizing output vector for the chained form as

$$z = \begin{pmatrix} x_1 \\ x_3 \\ x_5 \\ x_6 \end{pmatrix} \tag{4.24}$$

Differentiating with respect to time we get

$$\dot{z} = \begin{pmatrix} \dot{x}_1 \\ \dot{x}_3 \\ \dot{x}_5 \\ \dot{x}_6 \end{pmatrix} = \begin{pmatrix} 1 & 0 & 0 & 0 \\ x_2 & 0 & 0 & 0 \\ x_4 & 0 & 0 & 0 \\ 0 & 0 & 0 & 1 \end{pmatrix} \begin{pmatrix} u_1 \\ u_2 \\ u_3 \\ u_4 \end{pmatrix} \tag{4.25}$$

In order for the algorithm to proceed, we need to add two integrators whose states are denoted by ζ_1 and ζ_2 on the inputs u_1 and u_4, respectively, so their differentiation in the next step is avoided. Thus,

$$\begin{aligned} u_1 &= \zeta_1 \quad u_4 = \zeta_2 \\ \dot{\zeta}_1 &= u_1' \quad \dot{\zeta}_2 = u_4' \end{aligned} \tag{4.26}$$

where u_1' and u_4' are the new auxiliary inputs on the system. Substituting the above values and differentiating the output vector again we obtain

$$\ddot{z} = \begin{pmatrix} \dot{u}_1 \\ \dot{x}_2 u_1 + x_2 \dot{u}_1 \\ \dot{x}_4 u_1 + x_4 \dot{u}_1 \\ \dot{u}_4 \end{pmatrix} = \begin{pmatrix} \dot{\zeta}_1 \\ \dot{x}_2 \zeta_1 + x_2 \dot{\zeta}_1 \\ \dot{x}_4 u_1 + x_4 \dot{\zeta}_1 \\ \dot{\zeta}_2 \end{pmatrix}$$

or

$$\ddot{z} = \begin{pmatrix} u_1' \\ \zeta_1 u_2 + x_2 u_1' \\ \zeta_1 u_3 + x_4 u_1' \\ u_4' \end{pmatrix} = \begin{pmatrix} 1 & 0 & 0 & 0 \\ x_2 & \zeta_1 & 0 & 0 \\ x_4 & 0 & \zeta_1 & 0 \\ 0 & 0 & 0 & 1 \end{pmatrix} \begin{pmatrix} u_1' \\ u_2 \\ u_3 \\ u_4' \end{pmatrix} \tag{4.27}$$

In the above equations all the inputs appear in a nonsingular way, that is, the decoupling matrix is nonsingular. The value of the determinant of the matrix is ζ_1^2. Thus, the algorithm terminates after two differentiations. The matrix is nonsingular only as long as $\zeta_1 \neq 0$ or $u_1 \neq 0$. Here the order of the compensator is two ($b = 2$) and the number of states in the system is six ($n = 6$). The sum of the order of differentiations that is eight is equal to $n + b$. Thus, full-state linearization is achieved.

Let Equation 4.27 be rewritten as $\ddot{z} = r$, where r is the auxiliary reference input. Therefore, we have the following decoupled chains of integrators

$$\begin{aligned} \ddot{z}_1 &= r_1 \\ \ddot{z}_2 &= r_2 \\ \ddot{z}_3 &= r_3 \\ \ddot{z}_4 &= r_4 \end{aligned} \tag{4.28}$$

The resulting nonlinear dynamic feedback controller is

$$\begin{aligned} u_1 &= \zeta_1 \\ u_2 &= (r_2 - x_2 r_1)/\zeta_1 \\ u_3 &= (r_3 - x_4 r_1)/\zeta_1 \\ u_4 &= \zeta_2 \\ \dot{\zeta}_1 &= u_1' = r_1 \\ \dot{\zeta}_2 &= u_4' = r_4 \end{aligned} \tag{4.29}$$

Assuming the system follows a smooth desired reference trajectory in chained-form coordinates as $z_d(t) = (x_{d1}(t), x_{d3}(t), x_{d5}(t), x_{d6}(t))$, the exponentially stabilizing feedback control law for the linear and decoupled system about this desired trajectory is given as

$$r_i = \ddot{z}_{di}(t) + k_{vi}(\dot{z}_{di}(t) - \dot{z}_i(t)) + k_{pi}(z_{di}(t) - z_i(t)); \quad i = 1,2...,4 \tag{4.30}$$

where k_{vi} and k_{pi} are the proportional and derivative (PD) gains and are chosen such that they are positive and the characteristic polynomials

$$s^2 + k_{vi}s + k_{pi}; \quad i = 1,2...,4 \tag{4.31}$$

are Hurwitz. The desired values for the variables $\dot{z}_d$ and $\ddot{z}_d$ are obtained from Equations 4.25 and 4.27. For the simulation the other state variables can be expressed in terms of the desired output trajectory at the initial time $t = t_0$:

$$\begin{aligned}
x_1(t_0) &= z_{d1}(t_0) = x_d(t_0) \\
x_2(t_0) &= \dot{z}_{d2}(t_0)/\dot{z}_{d1}(t_0) = y_d(t_0)/\dot{x}_d(t_0) \\
x_3(t_0) &= z_{d2}(t_0) = y_d(t_0) \\
x_4(t_0) &= \dot{z}_{d3}(t_0)/\dot{z}_{d1}(t_0) = \dot{z}_d(t_0)/\dot{x}_d(t_0) \\
x_{d5}(t_0) &= z_{d3}(t_0) = z_d(t_0) \\
x_{d6}(t_0) &= z_{d4}(t_0) \\
\zeta_1(t_0) &= \dot{z}_{d1}(t_0) \\
\zeta_2(t_0) &= \dot{z}_{d4}(t_0)
\end{aligned} \tag{4.32}$$

From this initialization, the output trajectory is reproducible. Any other initialization gives tracking error, which exponentially goes to zero with time. The same results will follow if the dynamic extension is applied to the original kinematic equations.

4.4.3 Simulation of the Controller

For simulation, here again the same desired trajectory is used as was used in the linearized case. The trajectory chosen is

$$\begin{aligned}
z_{d1}(t) &= x_d(t) = t \quad & z_{d2}(t) &= y_d(t) = a\sin\omega t \\
z_{d3}(t) &= z_d(t) = 1 \quad & z_{d4}(t) &= x_{d6}(t) = 0
\end{aligned}$$

which gives the desired values for the chained-form states and inputs as follows:

$$x_{d1}(t) = t \quad x_{d2}(t) = a\omega\cos\omega t \quad x_{d3}(t) = a\sin\omega t$$

$$x_{d4}(t) = 0 \quad x_{d5}(t) = 1 \quad x_{d6}(t) = 0$$

and

$$\zeta_{d1}(t) = u_{d1}(t) = 1 \quad \zeta_{d2}(t) = u_{d4}(t) = 0$$

$$u_{d2}(t) = -a\omega^2\sin\omega t \quad u_{d3}(t) = 0$$

The initial conditions for the states are

$$x_{d1}(0) = 0 \quad x_{d2}(0) = a\omega \quad x_{d3}(0) = 0$$

$$x_{d4}(0) = 0 \quad x_{d5}(0) = 1 \quad x_{d6}(0) = 0$$

$$\zeta_{d1}(0) = u_{d1}(0) = 1 \quad \zeta_{d2}(0) = u_{d4}(0) = 0$$

Choosing again the six coincident closed-loop poles at –2, that is, $p_i = -2$; $i = 1,\ldots,6$, we get the PD gains as $k_{vi} = 4$ and $k_{pi} = 4$. Choosing $a = 1$ and $\omega = \pi$ we get the simulation results. The results show the tracking errors for chained-form states and inputs, and for actual states and inputs. Once the tracking errors go to zero, the actual inputs (control inputs) are the same as the desired inputs. The simulation results are presented in Figures 4.13 through 4.24.

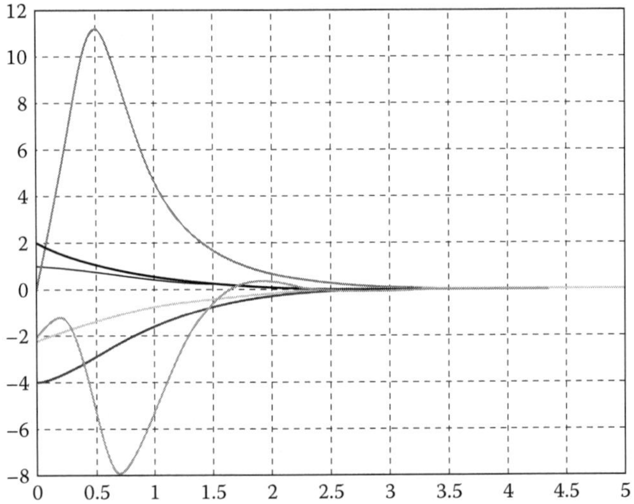

FIGURE 4.13 The result of dynamic feedback: tracking errors in chained-form variables vs. time (s).

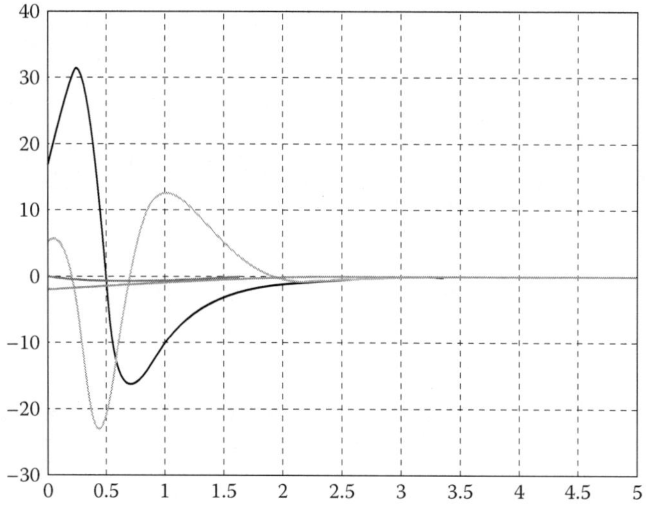

FIGURE 4.14 The result of dynamic feedback: tracking errors in chained-form inputs vs. time (s).

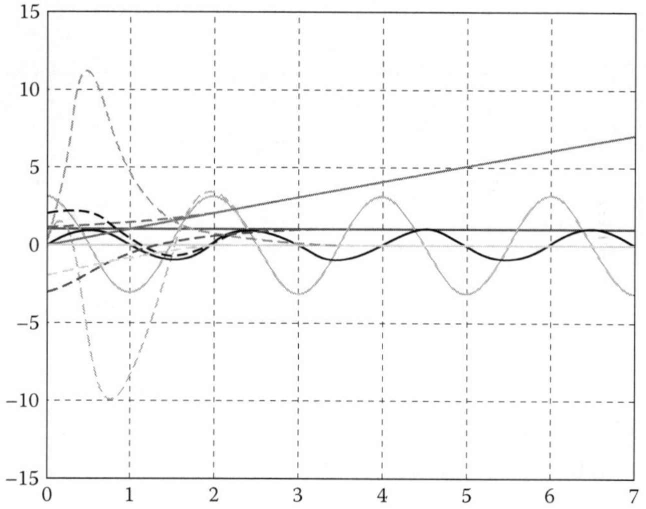

FIGURE 4.15 The result of dynamic feedback: actual (--) and desired (-) chained-form variables vs. time (s).

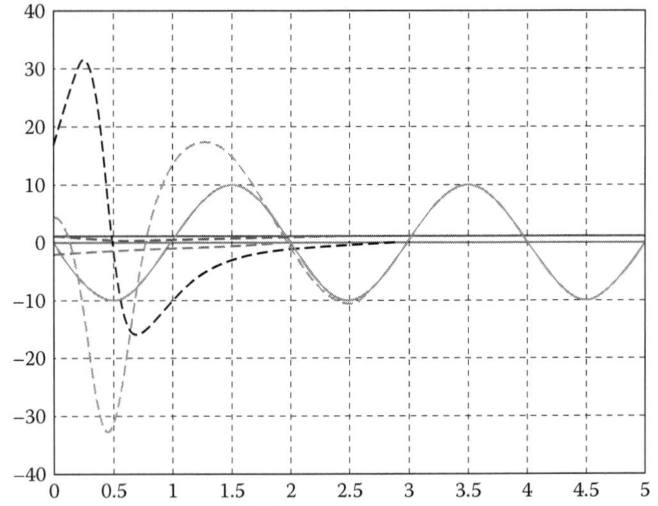

FIGURE 4.16 The result of dynamic feedback: actual (--) and desired (-) chained-form inputs vs. time (s).

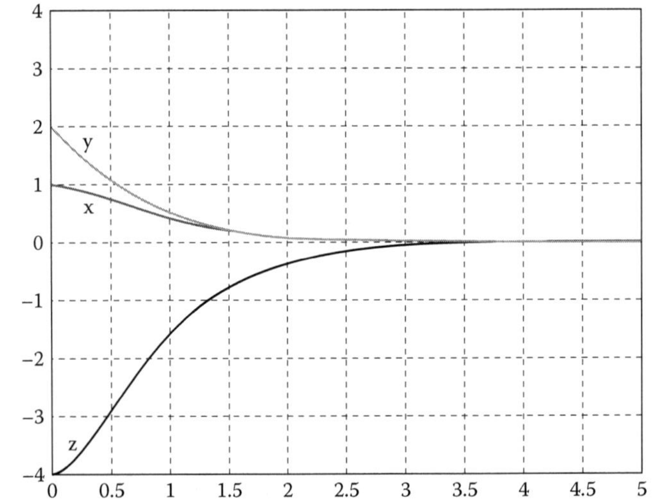

FIGURE 4.17 The result of dynamic feedback: tracking errors (m) in variables *x*, *y*, and *z* vs. time (s).

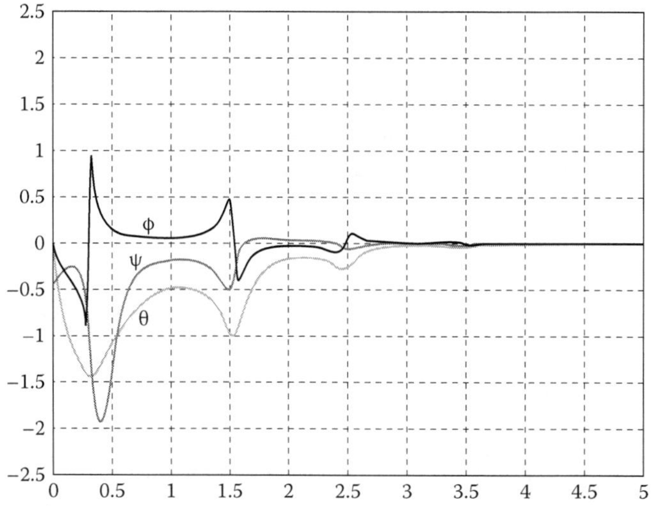

FIGURE 4.18 The result of dynamic feedback: tracking errors (rad) in variables ψ, θ, and φ vs. time (s).

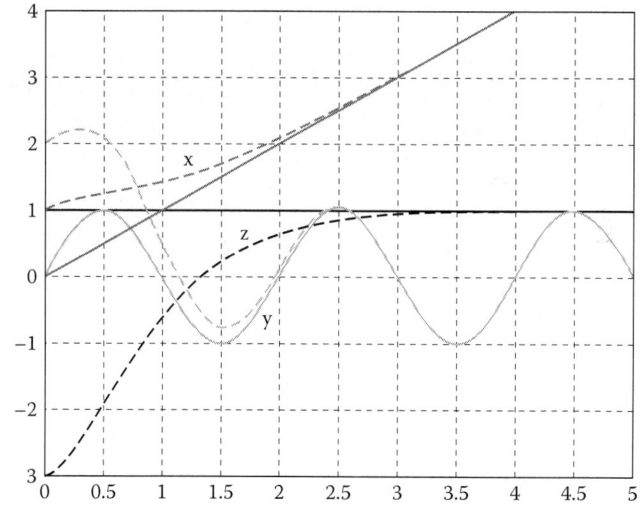

FIGURE 4.19 The result of dynamic feedback: actual (--) and desired (-) original variables *x*, *y*, *z* vs. time (*s*).

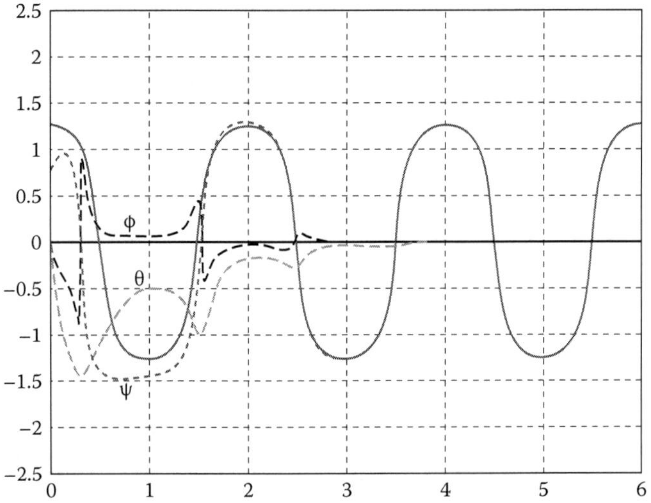

FIGURE 4.20 The result of dynamic feedback: actual (--) and desired (-) original variables ψ, θ, and φ vs. time (s).

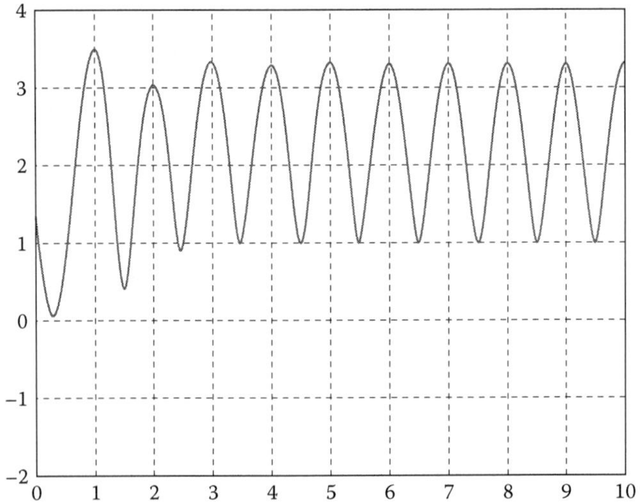

FIGURE 4.21 The result of dynamic feedback: v_1 (m/s) vs. time (s).

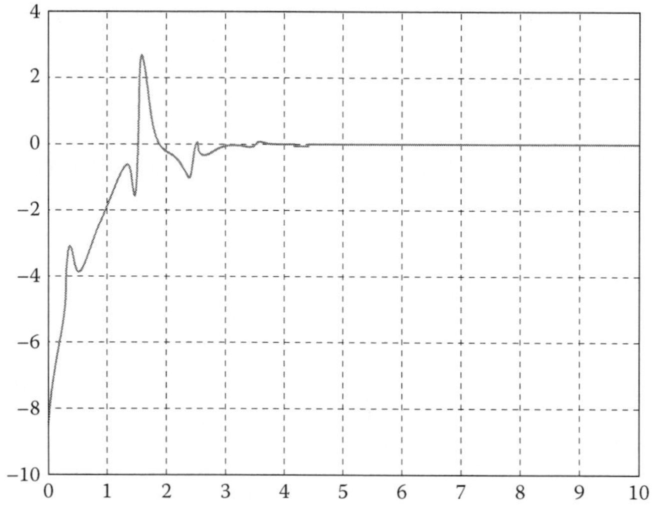

FIGURE 4.22 The result of dynamic feedback: v_2 (rad/s) vs. time (s).

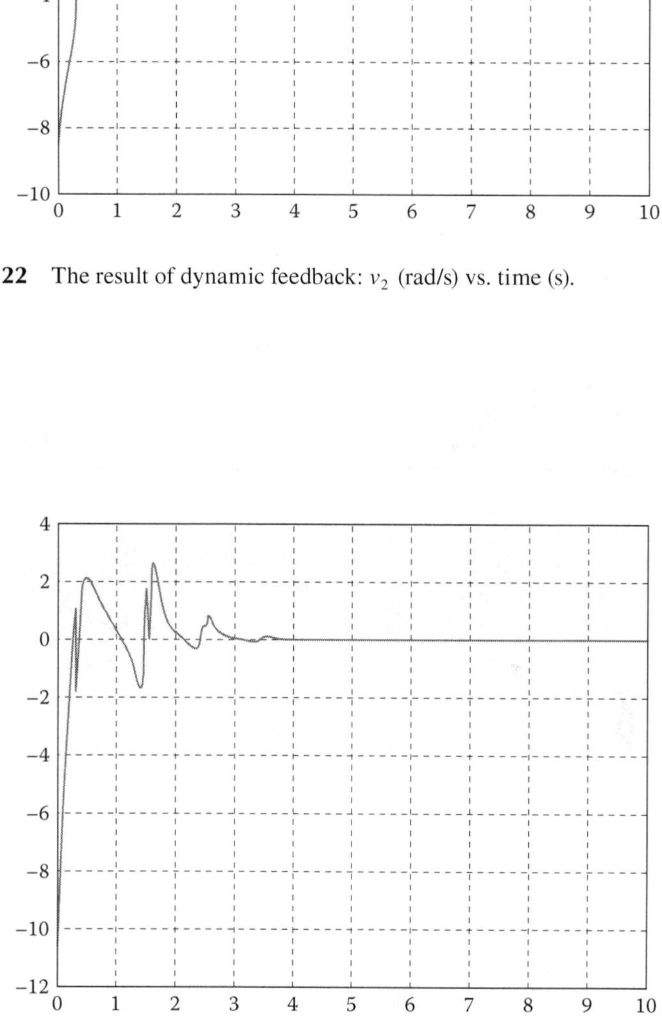

FIGURE 4.23 The result of dynamic feedback: v_3 (rad/s) vs. time (s).

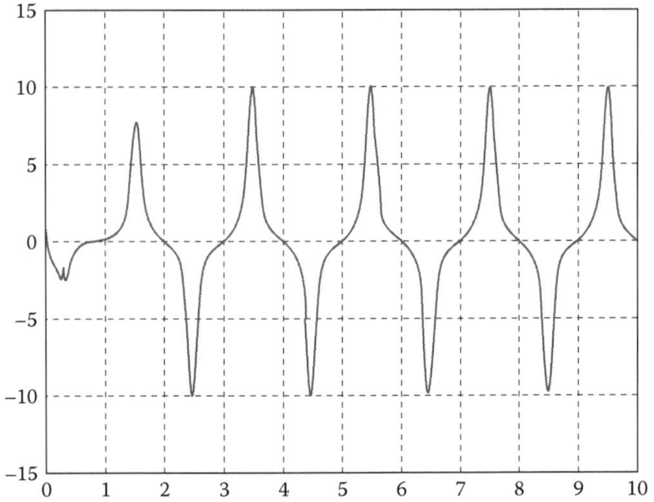

FIGURE 4.24 The result of dynamic feedback: v_4 (rad/s) vs. time (s).

4.4.4 MATLAB Program Code for Dynamic Extension

```
%%%%%%%%%%%%%%%%%%%%%%%%%%%%%%%%%%%%%%
% Approximate linearization through Dynamic Extension %
%%%%%%%%%%%%%%%%%%%%%%%%%%%%%%%%%%%%%%
t=[0:0.01:10]'; a=1; w=pi;
xd=t; xddot=1;
yddot=a*w*cos(w*t);
y2ddot=-a*w*w*sin(w*t); yd=a*sin(w*t);
den=(sqrt(xddot.*xddot+yddot.*yddot)).*(sqrt(xddot.*xddot
+yddot.*yddot)).*(sqrt(xddot.*xddot+yddot.*yddot));
     [t,x]=ode45('F1',[t],[-2,1]);
     [t,y]=ode45('F2',[t],[a*w,1]);
     [t,z]=ode45('F3',[t],[0,1]);
xe=-x(:,1)+xd;
ye=-y(:,1)+yd;
ze=-z(:,1)+zd;
theta=atan(y(:,2)./x(:,2));
thetad=atan(a*w*cos(w*t));
thetae=thetad-theta;
x2=(x(:,2).*y(:,3)-x(:,3).*y(:,2))./
(x(:,1).*x(:,1).*x(:,1));
phi=atan(cos(theta).*cos(theta).*cos(theta).*x2);
phid=atan((y2ddot.*xddot)./den);
phie=phid-phi;
%%%%%%%%%%%%%%%%%%%%%%%%%%%%
```

```
% Matlab Functions 'F1', 'F2', 'F3' and 'F4'%
%%%%%%%%%%%%%%%%%%%%%%%%%
function xp = F(t,x)    %%% function 'F1'
a=1; w=pi;
xp = zeros(2,1);
xp(1)=x(2);
xp(2)=-4*x(1)-4*x(2)+4+4*t;
function yp = F (t,y) %%% function 'F2'
a=1; w=pi;
yp = zeros(2,1);
yp(1)=y(2);
yp(2)=-4*y(2)-4*y(1)-a*w*sin(w*t)+4*a*w*cos(w*t)+4*a*sin
(w*t);
function zp = F4(t,z)    %%% function 'F3'
a=1; w=pi;
zp = zeros(2,1);
zp(1)=z(2);
zp(2)=-4*z(2)-4*z(1)+4;
function sp = F4(t,s)    %%% function 'F4'
a=1; w=pi;
sp = zeros(2,1);
sp(1)=s(2);
sp(2)=-4*s(2)-4*s(1);
%%%%%%%%%%%%%%%%%%%%%%%%%%%%%%%%%%%%%%%%%%
% Tracking errors in chained form variables %
%%%%%%%%%%%%%%%%%%%%%%%%%%%%%%%%%%%%%%%%%%
t=[0:0.001:9.999]'; a=1; w=pi;
xd=t;
yd=a*sin(w*t);
zd=1; sd=0;
xd2=a*w*cos(w*t);
xd4=0*ones(1,10000);
PSId=atan(xd2);
thetad=-atan(xd4'./sec(PSId));
phid=acot(cot(thetad)./sin(PSId)+tan(PSId)./sin(thetad)) ;
xddot=1;
yddot=a*w*cos(w*t);
y2ddot=-a*w*w*sin(w*t); yd=a*sin(w*t);
den=(sqrt(xddot.*xddot+yddot.*yddot)).*(sqrt(xddot.*xddo
t+yddot.*yddot)).*(sqrt(xddot.*xddot+yddot.*yddot));
% %% response for the chained form error variables %%%
     [t,x]=ode45('F1',[t],[1,1]);
     [t,y]=ode45('F2',[t],[2,1]);
     [t,z]=ode45('F3',[t],[-3,0]);
     [t,s]=ode45('F4',[t],[-2,0]);
```

```
x2=y(:,2)./x(:,2);
x4=z(:,2)./x(:,2);
xe=x(:,1)-xd;
ye=y(:,1)-yd;
ze=z(:,1)-zd';
se=s(:,1)-sd;
xe2=x2-xd2;
xe4=x4-xd4';
Figure (1)
plot(t, xe,'b'), grid, title('Error in chained form
states'); hold on
plot(t, xe2,'g');
plot(t,ye,'r');
plot(t,xe4,'c');
plot(t, ze,'m')
plot(t, se,'y'); hold off
%%%%%%%%%%%%%%%%%%%%%%%%%%%%%%%%%%%%%
% Computation of tracking errors in chained form inputs %
%%%%%%%%%%%%%%%%%%%%%%%%%%%%%%%%%%%%%
xd=t;   yd=a*sin(w*t); zd=1; t=[0:0.001:9.999]';
xddot=1;
yddot=a*w*cos(w*t);
y2ddot=-a*w*w*sin(w*t); yd=a*sin(w*t);
den=(sqrt(xddot.*xddot+yddot.*yddot)).*(sqrt(xddot.*xddo
t+yddot.*yddot)).*(sqrt(xddot.*xddot+yddot.*yddot));
     [t,x]=ode45('F1',[t],[1,1]);
     [t,y]=ode45('F2',[t],[2,1]);
     [t,z]=ode45('F3',[t],[-3,0]);
% %% error in chained form inputs %%%
r1=4+4*t-4*x(:,2)-4*x(:,1);
r2=-4*y(:,2)-4*y(:,1)-a*w*w*sin(w*t)+4*a*w*cos(w*t)+4*a*
sin(w*t);
r3=-4*z(:,2)-4*z(:,1)+4;
u1=x(:,2);
zeta1=u1;
u2=(r2-x2.*r1)./zeta1;
u3=(r3-x4.*r1)./zeta1;
u4=s(:,1);
ud1=1*ones(1,10000);
u1e=u1-ud1';
ud2=-a*w^2*sin(w*t);
u2e=u2-ud2;
ud3=0*ones(1,10000);
u3e=u3-ud3';
ud4=0*ones(1,10000);
u4e=u4-ud4';
```

```
Figure (2)
plot(t,u1e,'b'), grid, title('Error in chained inputs');
hold on
plot (t, u2e,'g');
plot(t,u3e,'r');
plot(t,u4e,'c'); hold off
%%%%%%%%%%%%%%%%%%%%%%%%%%
% Actual and desired chained form variables %
%%%%%%%%%%%%%%%%%%%%%%%%%%
t=[0:0.001:9.999]'; a=1; w=pi;
xd=t; yd=a*sin(w*t);
zd=1; sd=0;
xd2=a*w*cos(w*t);
xd4=0*ones(1,10000);
PSId=atan(xd2);
thetad=-atan(xd4'./sec(PSId));
phid=acot(cot(thetad)./sin(PSId)+tan(PSId)./sin(thetad)) ;
xddot=1;
yddot=a*w*cos(w*t);
y2ddot=-a*w*w*sin(w*t); yd=a*sin(w*t);
den=(sqrt(xddot.*xddot+yddot.*yddot)).*(sqrt(xddot.*xddo
t+yddot.*yddot)).*(sqrt(xddot.*xddot+yddot.*yddot));
%%% response for the chained form actual variables %%%
      [t,x]=ode45('F1',[t],[1,1]);
      [t,y]=ode45('F2',[t],[2,1]);
      [t,z]=ode45('F3',[t],[-3,0]);
      [t,s]=ode45('F4',[t],[-2,0]);
x2=y(:,2)./x(:,2);
x4=z(:,2)./x(:,2);
Figure (3)
plot(t, xd,'b'), grid, title (' Chained form desired and
actual variables ');hold on
plot(t, x(:,1),'--');
plot(t, xd2,'g');
plot(t, x2,'g--');
plot(t,yd,'r');
plot(t,y(:,1),'r--');
plot(t,xd4,'c');
plot(t, x4,'c--');
plot(t, zd,'m');
plot(t, z(:,1),'m--');
plot(t, sd,'y')
plot(t, s(:,1),'y--'); hold off

%%%%%%%%%%%%%%%%%%%%%%%%%%%%%%%%
% Computation of actual and desired chained form inputs %
```

```
%%%%%%%%%%%%%%%%%%%%%%%%%%%%%%
t=[0:0.001:9.999]'; a=1; w=pi;
     [t,x]=ode45('F1',[t],[1,1]);
     [t,y]=ode45('F2',[t],[2,1]);
     [t,z]=ode45('F3',[t],[-3,0]);
     [t,s]=ode45('F4',[t],[-2,0]);
x2=y(:,2)./x(:,2);
x4=z(:,2)./x(:,2);
%%% response for the chained form actual and desired
inputs%%%
r1=4+4*t-4*x(:,2)-4*x(:,1);
r2=-4*y(:,2)-4*y(:,1)-a*w*w*sin(w*t)+4*a*w*cos(w*t)+4*a*
sin(w*t);
r3=-4*z(:,2)-4*z(:,1)+4;
u1=x(:,2);
zeta1=u1;
u2=(r2-x2.*r1)./zeta1;
u3=(r3-x4.*r1)./zeta1;
u4=s(:,1);
ud1=1*ones(1,10000);
ud2=-a*w^2*sin(w*t);
ud3=0*ones(1,10000);
ud4=0*ones(1,10000);
Figure (4)
plot(t, ud1,'b'), grid, title ('Chained form desired and
actual inputs'); hold on
plot(t, u1, '--');
plot(t, ud2,'g');
plot(t,u2,'g--');
plot(t,ud3,'r');
plot(t,u3,'r--');
plot(t,ud4,'c');
plot(t, u4,'c--'); hold off
%%%%%%%%%%%%%%%%%%%%%%%%%%%%%
% Computation of tracking errors in variables x , y , z %
%%%%%%%%%%%%%%%%%%%%%%%%%%%%%
t=[0:0.001:9.999]'; a=1; w=pi;
xd=t;
yd=a*sin(w*t);
zd=1;
xddot=1;
yddot=a*w*cos(w*t);
y2ddot=-a*w*w*sin(w*t); yd=a*sin(w*t);
den=(sqrt(xddot.*xddot+yddot.*yddot)).*(sqrt(xddot.*xddo
t+yddot.*yddot)).*(sqrt(xddot.*xddot+yddot.*yddot));
```

```
     [t,x]=ode45('F1',[t],[1,1]);
     [t,y]=ode45('F2',[t],[2,1]);
     [t,z]=ode45('F3',[t],[-3,0]);
%%%response for the errors in actual system variables%%%
xe=x(:,1)-xd;
ye=y(:,1)-yd;
ze=z(:,1)-zd';
Figure (5)
plot(t, xe,'b'), grid,title('Error in original state
variables, x,y,z '); grid on
plot(t, ye,'g');
plot(t,ze,'r'); hold off
%%%%%%%%%%%%%%%%%%%%%%%%%%%%%%%%%
% Computation of tracking errors in variables ψ, θ, φ %
%%%%%%%%%%%%%%%%%%%%%%%%%%%%%%%%%
xd=t;
yd=a*sin(w*t);
zd=1;
sd=0;
xd2=a*w*cos(w*t);
xd4=0*ones(1,10000);
PSId=atan(xd2);
thetad=-atan(xd4'./sec(PSId));
phid=acot(cot(thetad)./sin(PSId)+tan(PSId)./sin(thetad)) ;
xddot=1;
yddot=a*w*cos(w*t);
y2ddot=-a*w*w*sin(w*t); yd=a*sin(w*t);
den=(sqrt(xddot.*xddot+yddot.*yddot)).*(sqrt(xddot.*xddo
t+yddot.*yddot)).*(sqrt(xddot.*xddot+yddot.*yddot));
     [t,x]=ode45('F1',[t],[1,1]);
     [t,y]=ode45('F2',[t],[2,1]);
     [t,z]=ode45('F3',[t],[-3,0]);
x2=y(:,2)./x(:,2);
x4=z(:,2)./x(:,2);
%%% response for the errors in actual system variables
(psi, theta, phi)%%%
PSI=atan(x2);
PSIe=PSI-PSId;
theta=-atan(x4./sec(PSI));
thetae=theta-thetad;
phi=acot(cot(theta)./sin(PSI)+tan(PSI)./sin(theta));
phie=phi-phid;
Figure (6)
plot(t, PSIe, 'b');%, grid, title('Error in original
state variables,(psi, theta, phi)');hold on
```

```
plot(t, thetae,' g')
plot(t, phie,' r'); hold off
%%%%%%%%%%%%%%%%%%%%%%%%%%%%
% Actual and desired original variables x, y, z %
%%%%%%%%%%%%%%%%%%%%%%%%%%%%
t=[0:0.001:9.999]'; a=1; w=pi;
xd=t;
yd=a*sin(w*t);
zd=1;
xddot=1;
yddot=a*w*cos(w*t);
y2ddot=-a*w*w*sin(w*t); yd=a*sin(w*t);
den=(sqrt(xddot.*xddot+yddot.*yddot)).*(sqrt(xddot.*xddo
t+yddot.*yddot)).*(sqrt(xddot.*xddot+yddot.*yddot));
     [t,x]=ode45('F1',[t],[1,1]);
     [t,y]=ode45('F2',[t],[2,1]);
     [t,z]=ode45('F3',[t],[-3,0]);
x2=y(:,2)./x(:,2);
x4=z(:,2)./x(:,2);
Figure (7)
plot(t, xd,'b'), grid, title('Original system actual and
desired state variables,(x, y, z)';hold on
plot(t, x(:,1),'--');
plot(t, yd,'g');
plot(t, y(:,1),'g--');
plot(t,zd,'r');
plot(t,z(:,1),'r--'); hold off
%%%%%%%%%%%%%%%%%%%%%%
% Actual and desired original variables ψ, θ, φ %
%%%%%%%%%%%%%%%%%%%%%%
xd=t;
yd=a*sin(w*t);
zd=1;
xddot=1;
yddot=a*w*cos(w*t);
y2ddot=-a*w*w*sin(w*t); yd=a*sin(w*t);
den=(sqrt(xddot.*xddot+yddot.*yddot)).*(sqrt(xddot.*xddo
t+yddot.*yddot)).*(sqrt(xddot.*xddot+yddot.*yddot));
xd2=a*w*cos(w*t);
xd4=0*ones(1,10000);
PSId=atan(xd2);
thetad=-atan(xd4'./sec(PSId));
phid=acot(cot(thetad)./sin(PSId)+tan(PSId)./sin(thetad)) ;
     [t,x]=ode45('F1',[t],[1,1]);
     [t,y]=ode45('F2',[t],[2,1]);
```

```
      [t,z]=ode45('F3',[t],[-3,0]);
      [t,s]=ode45('F4',[t],[-2,0]);
x2=y(:,2)./x(:,2);
x4=z(:,2)./x(:,2);
PSI=atan(x2);
PSIe=PSI-PSId;
theta=-atan(x4./sec(PSI));
thetae=theta-thetad;
phi=acot(cot(theta)./sin(PSI)+tan(PSI)./sin(theta));
phie=phi-phid;
Figure (8)
plot(t, PSId, 'b');%, grid, title('Original system
actual state variables, (psi, theta, phi) '); hold on
plot(t,PSI,'--');
plot(t, thetad, 'g')
plot(t,theta,' g--');
plot(t, phid, 'r');
plot(t, phi,'r--'); hold off

%%%%%%%%%%%%%%%%%%%%%%
% Computation of actual input v1%
%%%%%%%%%%%%%%%%%%%%%%
xd=t;
yd=a*sin(w*t);
zd=1;
xddot=1;
yddot=a*w*cos(w*t);
y2ddot=-a*w*w*sin(w*t); yd=a*sin(w*t);
den=(sqrt(xddot.*xddot+yddot.*yddot)).*(sqrt(xddot.*xddo
t+yddot.*yddot)).*(sqrt(xddot.*xddot+yddot.*yddot));
      [t,x]=ode45('F1',[t],[1,1]);
      [t,y]=ode45('F2',[t],[2,1]);
      [t,z]=ode45('F3',[t],[-3,0]);
      [t,s]=ode45('F4',[t],[-2,0]);
x2=y(:,2)./x(:,2);
x4=z(:,2)./x(:,2);
PSI=atan(x2);
theta=-atan(x4./sec(PSI));
phi=acot(cot(theta)./sin(PSI)+tan(PSI)./sin(theta));
%%% Actual control variable v1 %%%
r1=4+4*t-4*x(:,2)-4*x(:,1);
r2=-4*y(:,2)-4*y(:,1)-a*w*w*sin(w*t)+4*a*w*cos(w*t)+4*a*
sin(w*t);
r3=-4*z(:,2)-4*z(:,1)+4;
u1=x(:,2);
```

```
zeta1=u1;
u2=(r2-x2.*r1)./zeta1;
u3=(r3-x4.*r1)./zeta1;
u4=s(:,1);
v1=u1./cos(PSI).*cos(theta);
Figure (9);
plot(t,v1), grid, title( 'original system input1 (v1');
%%%%%%%%%%%%%%%%%%%%%%%%%%
% Computation of input v2 %
%%%%%%%%%%%%%%%%%%%%%%%%%%%
xd=t;
yd=a*sin(w*t);
zd=1;
xddot=1;
yddot=a*w*cos(w*t);
y2ddot=-a*w*w*sin(w*t); yd=a*sin(w*t);
den=(sqrt(xddot.*xddot+yddot.*yddot)).*(sqrt(xddot.*xddo
t+yddot.*yddot)).*(sqrt(xddot.*xddot+yddot.*yddot));
     [t,x]=ode45('F1',[t],[1,1]);
     [t,y]=ode45('F2',[t],[2,1]);
     [t,z]=ode45('F3',[t],[-3,0]);
     [t,s]=ode45('F4',[t],[-2,0]);
x2=y(:,2)./x(:,2);
x4=z(:,2)./x(:,2);
PSI=atan(x2);
theta=-atan(x4./sec(PSI));
phi=acot(cot(theta)./sin(PSI)+tan(PSI)./sin(theta));
r1=4+4*t-4*x(:,2)-4*x(:,1);
r2=-4*y(:,2)-4*y(:,1)-a*w*w*sin(w*t)+4*a*w*cos(w*t)+4*a*
sin(w*t);
r3=-4*z(:,2)-4*z(:,1)+4;
u1=x(:,2);
zeta1=u1;
u2=(r2-x2.*r1)./zeta1;
u3=(r3-x4.*r1)./zeta1;
u4=s(:,1);
%%% Actual control variable v2 %%%
v1=u1./cos(PSI).*cos(theta);
r11=cos(PSI).*cos(theta);
r23=sin(PSI).*sin(theta).*cos(phi)-cos(PSI).*sin(phi);
r33=cos(theta).*cos(phi);
r22=sin(PSI).*sin(theta).*sin(phi)+cos(PSI).*cos(phi);
r32=sin(phi).*cos(theta);
r12=cos(PSI).*sin(theta).*sin(phi)-sin(PSI).*cos(phi);
r13=sin(PSI).*sin(theta).*cos(phi)+sin(PSI).*sin(phi);
R=r11+r22+r33;
```

```
R1=1+R;
r11n=1./(1+r11);
v3=r11.*(-r23.*u2-r33.*u3);
v4=r11.*(r22.*u2+r32.*u3);
v2=r11n.*(R1.*u4-r12.*v3-r13.*v4);
Figure (10);
plot (t,v2),grid, title( 'original system input2 (v2');
%%%%%%%%%%%%%%%%%%%%%%%%%%
% Computation of input v3 %
%%%%%%%%%%%%%%%%%%%%%%%%%%
t=[0:0.001:9.999]';
xddot=1;
yddot=a*w*cos(w*t);
y2ddot=-a*w*w*sin(w*t); yd=a*sin(w*t);
den=(sqrt(xddot.*xddot+yddot.*yddot)).*(sqrt(xddot.*xddo
t+yddot.*yddot)).*(sqrt(xddot.*xddot+yddot.*yddot));
     [t,x]=ode45('F1',[t],[1,1]);
     [t,y]=ode45('F2',[t],[2,1]);
     [t,z]=ode45('F3',[t],[-3,0]);
     [t,s]=ode45('F4',[t],[-2,0]);
x2=y(:,2)./x(:,2);
x4=z(:,2)./x(:,2);
PSI=atan(x2);
theta=-atan(x4./sec(PSI));
phi=acot(cot(theta)./sin(PSI)+tan(PSI)./sin(theta));
r1=4+4*t-4*x(:,2)-4*x(:,1);
r2=-4*y(:,2)-4*y(:,1)-a*w*w*sin(w*t)+4*a*w*cos(w*t)+4*a*
sin(w*t);
r3=-4*z(:,2)-4*z(:,1)+4;
u1=x(:,2);
zeta1=u1;
u2=(r2-x2.*r1)./zeta1;
u3=(r3-x4.*r1)./zeta1;
u4=s(:,1);
%%% Actual control variable v3 %%%
r11=cos(PSI).*cos(theta);
r23=sin(PSI).*sin(theta).*cos(phi)-cos(PSI).*sin(phi);
r33=cos(theta).*cos(phi);
r22=sin(PSI).*sin(theta).*sin(phi)+cos(PSI).*cos(phi);
r32=sin(phi).*cos(theta);
r12=cos(PSI).*sin(theta).*sin(phi)-sin(PSI).*cos(phi);
r13=sin(PSI).*sin(theta).*cos(phi)+sin(PSI).*sin(phi);
R=r11+r22+r33;
R1=1+R;
r11n=1./(1+r11);
v3=r11.*(-r23.*u2-r33.*u3);
```

```
Figure (11);
plot(t,v3),grid, title( 'original system input3 (v3');
%%%%%%%%%%%%%%%%%%%%%%%%%%%
% Computation of input v4 %
%%%%%%%%%%%%%%%%%%%%%%%%%%%%
t=[0:0.001:9.999]';
     [t,x]=ode45('F1',[t],[1,1]);
     [t,y]=ode45('F2',[t],[2,1]);
     [t,z]=ode45('F3',[t],[-3,0]);
     [t,s]=ode45('F4',[t],[-2,0]);
x2=y(:,2)./x(:,2);
x4=z(:,2)./x(:,2);
PSI=atan(x2);
theta=-atan(x4./sec(PSI));
phi=acot(cot(theta)./sin(PSI)+tan(PSI)./sin(theta));
r1=4+4*t-4*x(:,2)-4*x(:,1);
r2=-4*y(:,2)-4*y(:,1)-a*w*w*sin(w*t)+4*a*w*cos(w*t)+4*a*
sin(w*t);
r3=-4*z(:,2)-4*z(:,1)+4;
u1=x(:,2);
zeta1=u1;
u2=(r2-x2.*r1)./zeta1;
u3=(r3-x4.*r1)./zeta1;
u4=s(:,1);
%%% Actual control variable v4 %%%
r11=cos(PSI).*cos(theta);
r23=sin(PSI).*sin(theta).*cos(phi)-cos(PSI).*sin(phi);
r33=cos(theta).*cos(phi);
r22=sin(PSI).*sin(theta).*sin(phi)+cos(PSI).*cos(phi);
r32=sin(phi).*cos(theta);
r12=cos(PSI).*sin(theta).*sin(phi)-sin(PSI).*cos(phi);
r13=sin(PSI).*sin(theta).*cos(phi)+sin(PSI).*sin(phi);
R=r11+r22+r33;
R1=1+R;
r11n=1./(1+r11);
v4=r11.*(r22.*u2+r32.*u3);
Figure (12);
plot(t,v4),grid, title('original system input4 (v4');
```

4.5 POINT-TO-POINT STABILIZATION

In the following section the problem of point-to-point stabilization is addressed. The system is supposed to reach a final desired configuration starting from an initial point, without the need to plan a trajectory. As stated earlier, point stabilization cannot be achieved by a smooth time-invariant feedback. Only nonsmooth or time-varying feedback laws are of interest for the task. For our system we will adopt the latter approach.

4.5.1 Control with Smooth Time-Varying Feedback

The method of designing a stabilizing control law here is the one proposed in [7]. The control law presented there was for a two-input nonholonomic system. The controller here is an extension of the same. The statement of the problem is: given a nonlinear drift-free control system (Equation 3.11)

$$\dot{q} = g_1(q)v_1 + g_2(q)v_2 + g_3(q)v_3 + g_4(q)v_4 \tag{4.33}$$

we have to find a control law of the form $v(q,t)$ that makes the origin globally stable. In [7] the origin of the control system (Equation 4.33) is stabilized as represented in power form. Thus, before going for the control design, we will convert Equation 4.33 into a power form.

4.5.2 Power Form

The method of converting Equation 4.33 to power form is presented in [24]. The transformation is done in two steps. In the first step the original system is converted into a three-chain, single-generator chained form, as described in Section 3.3. The chained form there obtained is given as

$$\begin{aligned} &\dot{x}_1 = x_1^0 = u_1 \quad \dot{x}_2 = x_2^0 = u_2 \quad \dot{x}_4 = x_3^0 = u_3 \\ &\dot{x}_6 = x_4^0 = u_4 \quad \dot{x}_3 = x_{20}^0 = x_2 u_1 \quad \dot{x}_5 = x_{31}^0 = x_4 u_1 \end{aligned} \tag{4.34}$$

In the second step the chained-form system (Equation 4.34) is converted into power form. For the three-chain single-generator chained form the global transformation to power form is given below [24]:

$$\begin{aligned} &y_j = x_j^0; 1 \le j \le 4 \\ &z_{j0}^k = (-1)^k x_{j0}^k + \sum_{n=0}^{k-1} (-1)^n \frac{1}{(k-n)!} \left(x_1^0\right)^{k-n} x_{j0}^n \; 2 \le j \le 4 \; 1 \le k \le n_j \end{aligned} \tag{4.35}$$

which gives the power form as

$$\begin{aligned} &\dot{y}_j = u_j; 1 \le j \le 4 \\ &z_{j0}^k = \frac{1}{k!}(y_1)^k u_j \; 2 \le j \le 4 \; 1 \le k \le n_j \end{aligned} \tag{4.36}$$

Here we should recall from Section 3.3 that $x_{j0}^0 = x_j^0$ is identified as the top of the chains and $\sum_{n=2}^{4} n_j = 2$. Thus, we have $n_2 = n_3 = 1$ and $n_4 = 0$. Using Equation 4.34 and taking the above values, the transformation Equation 4.35 becomes

$$
\begin{aligned}
y_1 &= x_1 \\
y_2 &= x_2 \\
y_3 &= x_4 \\
y_4 &= x_6 \\
z_1 &= z_{20}^1 = -x_3 + x_1 x_2 \\
z_2 &= z_{30}^1 = -x_5 + x_1 x_4 \\
z_3 &= z_{40}^0 = x_6
\end{aligned}
\tag{4.37}
$$

and the corresponding power form is

$$
\begin{aligned}
\dot{y}_1 &= u_1 \\
\dot{y}_2 &= u_2 \\
\dot{y}_3 &= u_3 \\
\dot{y}_4 &= u_4 \\
\dot{z}_{20}^1 &= y_1 u_2 \\
\dot{z}_{30}^1 &= y_1 u_3 \\
\dot{z}_{40}^0 &= u_4
\end{aligned}
\tag{4.38}
$$

4.5.3 Control Design with Smooth Time-Varying Feedback

The smooth time-varying feedback control law for Equation 4.38 as adopted from [7] is given here as

$$
\begin{aligned}
u_1 &= -y_1 + \sigma(\rho(z))(\cos t - \sin t) \\
u_2 &= -y_2 + c_1 z_1 \cos t \\
u_3 &= -y_3 + c_2 z_2 \cos t \\
u_4 &= -y_4
\end{aligned}
\tag{4.39}
$$

with c_1, $c_2 > 0$ and $\rho(z) = (z_1)^2 + (z_2)^2 + (z_3)^2$. The controls (Equation 4.39) asymptotically stabilize the origin of Equation 4.38. For global stabilization the saturation functions are introduced in the control law. These functions eliminate the destabilizing effects away from the origin. The control is thus given as

$$\begin{aligned} u_1 &= -y_1 + \sigma(\rho(z))(\cos t - \sin t) \\ u_2 &= -y_2 + c_1\sigma(z_1)\cos t \\ u_3 &= -y_3 + c_2\sigma(z_2)\cos t \\ u_4 &= -y_4 \end{aligned} \tag{4.40}$$

where $c_j > 0$ and $\sigma : R \rightarrow R$ is a nondecreasing C^3 saturation function with a magnitude less than some $\delta > 0$ and is linear between $(-\delta, \delta)$. For global stabilization δ should be small enough. The saturation function then satisfies the following:

1. $\sigma(z) = z$ when $|z| \leq \varepsilon$
2. $|\sigma(z)| \leq \delta$ for all z such that $0 < \varepsilon < \delta$

The closed-loop dynamics are given as

$$\begin{aligned} \dot{y}_1 &= -y_1 + \sigma(\rho(z))(\cos t - \sin t) \\ \dot{y}_2 &= -y_2 + c_1\sigma(z_1)\cos t \\ \dot{y}_3 &= -y_3 + c_2\sigma(z_2)\cos t \\ \dot{y}_4 &= -y_4 \\ \dot{z}_1 &= -y_1 y_2 + y_1 c_1 \sigma(z_1)\cos t \\ \dot{z}_2 &= -y_1 y_3 + y_1 c_2 \sigma(z_2)\cos t \end{aligned} \tag{4.41}$$

For some $0 < \varepsilon < \delta$, $\exists\delta_0$ such that if $\varepsilon < \varepsilon_0$, then the closed-loop dynamics are globally asymptotically stabilized to zero.

4.5.4 Simulation of the Controller

For simulation the value of δ is chosen to be 0.5 and the saturation functions $\sigma(\rho(z))$, $\sigma(z_1)$, and $\sigma(z_2)$ are 0.26, 0.5, and 0.1. The constants c_1 and c_2 are both chosen as 2. The initial values for y_1, y_2, y_3, and y_4 are chosen as –5, –2, –7, and –5, respectively. The simulation results are presented in Figures 4.25 through 4.34.

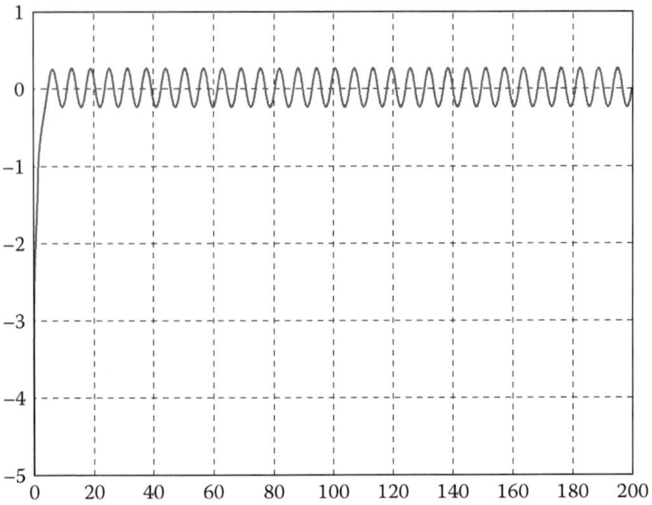

FIGURE 4.25 Point stabilization using time-varying feedback: x (m) vs. time (s).

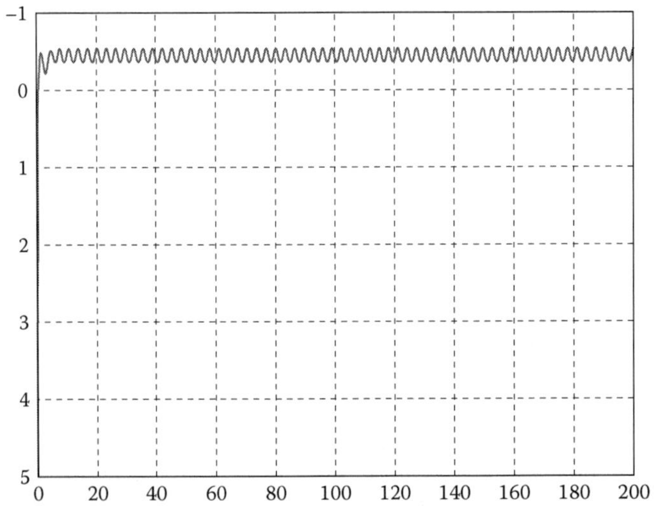

FIGURE 4.26 Point stabilization using time-varying feedback: y (rad) vs. time (s).

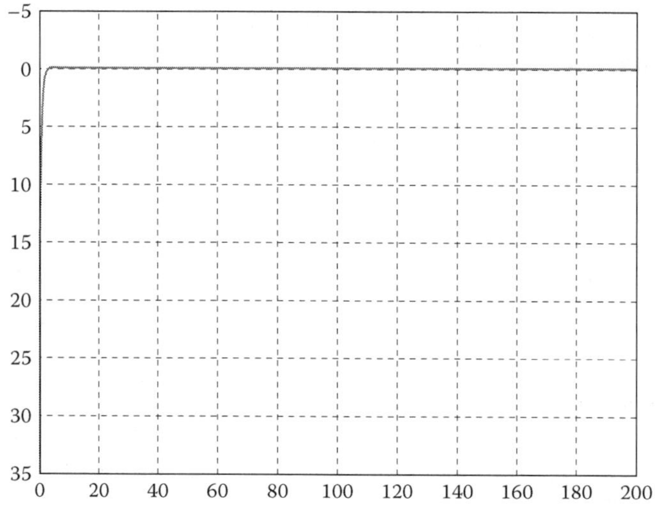

FIGURE 4.27 Point stabilization using time-varying feedback: z (m) vs. time (s).

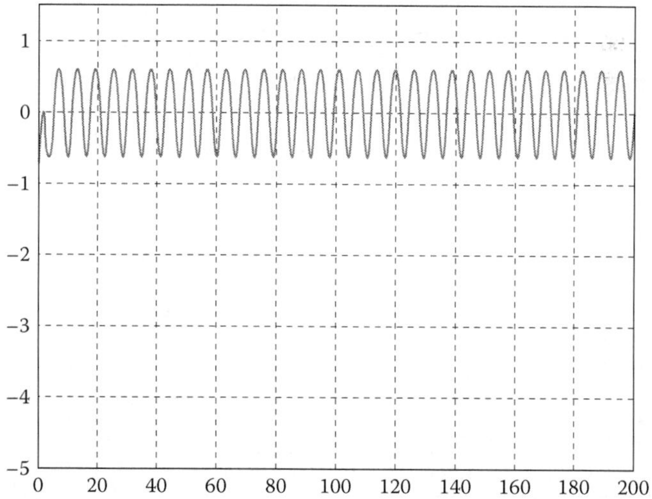

FIGURE 4.28 Point stabilization using time-varying feedback: ψ (rad) vs. time (s).

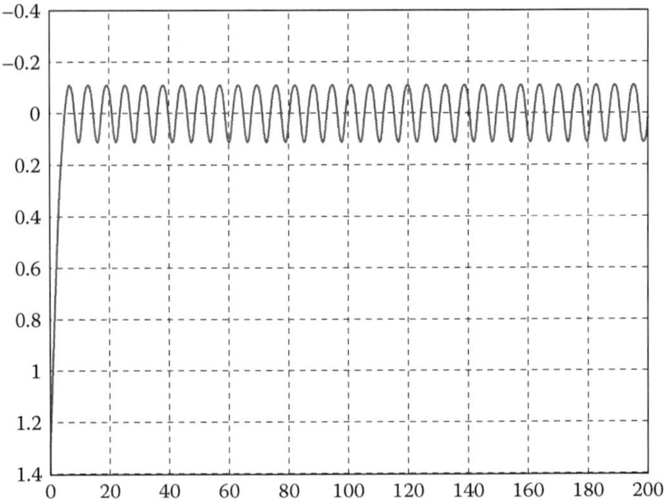

FIGURE 4.29 Point stabilization using time-varying feedback: θ (rad) vs. time (s).

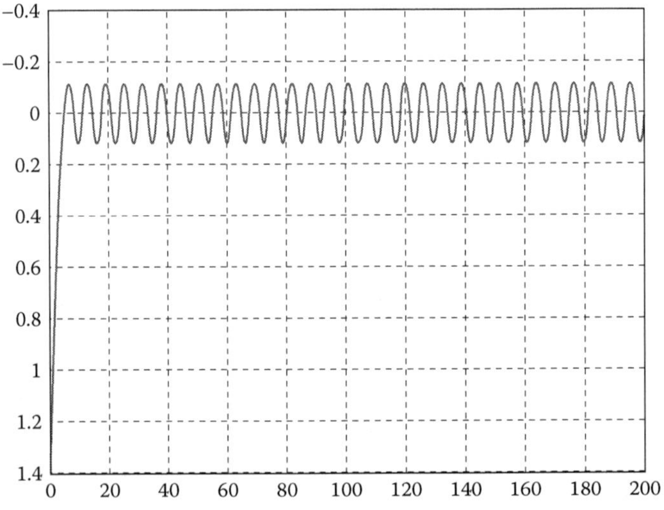

FIGURE 4.30 Point stabilization using time-varying feedback: ϕ (rad) vs. time (s).

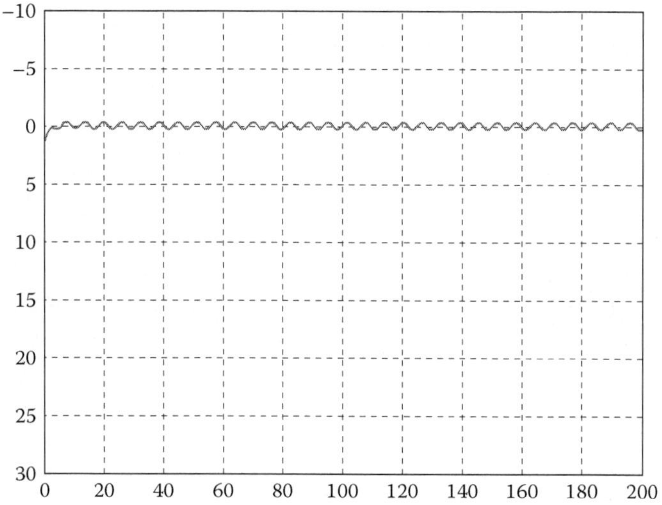

FIGURE 4.31 Point stabilization using time-varying feedback: v_1 (m/s) vs. time (s).

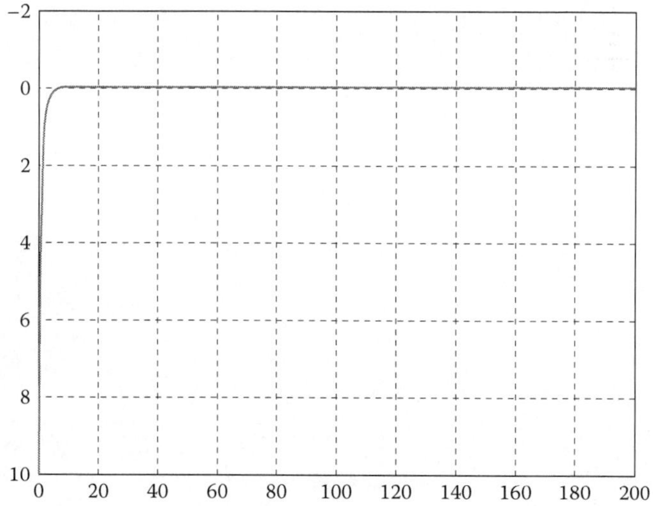

FIGURE 4.32 Point stabilization using time-varying feedback: v_2 (m/s) vs. time (s).

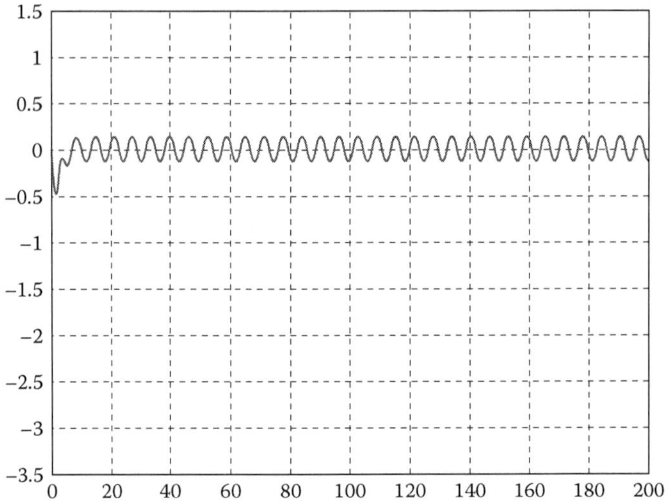

FIGURE 4.33 Point stabilization using time-varying feedback: v_3 (rad/s) vs. time (s).

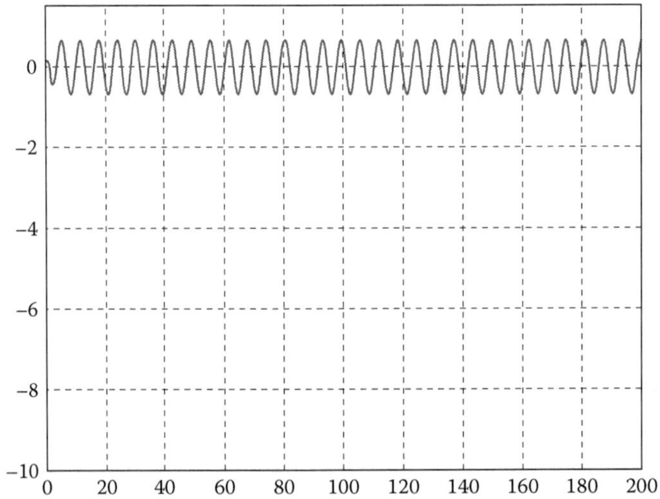

FIGURE 4.34 Point stabilization using time-varying feedback: v_4 (rad/s) vs. time (s).

4.5.5 MATLAB Program Code for Point Stabilization

```
%%%%%%%%%%%%%%%%%%%%%%%%%%%%%%%%%%%%%%
% Point Stabilization feedback controller %
%%%%%%%%%%%%%%%%%%%%%%%%%%%%%%%%%%%%%%
t=[0:0.01:120]; a=1; w=pi;
x=t;
xddot=1;
yd=a*sin(w*t);
zd=1*ones(1,1001);
yddot=a*w*cos(w*t);
y2ddot=-a*w*w*sin(w*t); yd=a*sin(w*t);
den=(sqrt(xddot.*xddot+yddot.*yddot)).*(sqrt(xddot.*xddo
t+yddot.*yddot)).*(sqrt(xddot.*xddot+yddot.*yddot));
     [t,y1]=ode45('yp1',[t],-5);
     [t,y2]=ode45('yp2',[t],-2);
     [t,y3]=ode45('yp3',[t],-7);
     [t,y4]=ode45('yp4',[t],-5);
x1=y1;
x2=y2;
x3=y1.*y2-0.5;
x4=y3;
x5=y1.*y3-0.1;
x6=y4;
%%%%%%%%%%%%%%%%%%%%%%%%%%%%%%
% Matlab Functions 'yp1', 'yp2', 'yp3' and 'yp4'%
%%%%%%%%%%%%%%%%%%%%%%%%%%%%%%
function y1 = yp1(t,y)    %%% function 'yp1'
a=1; w=pi;
y1 = zeros(3,1);
y (1)=y(2);
y (2)=y(3);
y (3)=-15*y(3)-75*y(2)-125*y(1)+75*a*w*cos(w*t)-
15*a*w*w*sin(w*t)-a*w*w*w*cos(w*t)+125*a*sin(w*t);
function y2 = yp2(t,y)    %%% function 'yp2'
a=1; w=pi;
y2 = zeros(3,1);
y (1)=y(2);
y (2)=y(3);
y (3)=-15*y(3)- 4*y(2)-4*y(1)+a*w*sin(w*t)+4*a*w*cos(w*t)
+4*a*sin(w*t);
function y3 = yp2(t,y)    %%% function 'yp3'
a=1; w=pi;
y3 = zeros(3,1);
```

```
y (1)=y(2);
y (2)=y(3);
y (3)=-4*y(3)-12*y(2)-4*y(1)+ 4*t+15*a*w*w*sin(w*t)-
a*w*w*w*cos(w*t)+125*a*sin(w*t);
function y4 = yp4(t,y)    %%% function 'yp4'
a=1; w=pi;
y4 = zeros(3,1);
y (1)=y(2);
y (2)=y(3);
y (3)=-10*y(3)-12*y(2)-130*y(1)+ a*w*sin(w*t)+
a*w*w*sin(w*t)-a*w*w*w*cos(w*t)+75*a*sin(w*t);
%%%%%%%%%%%%%%%%%%%%%%%%%%%%%%%%%%%%%%%%%%%
% Stabilization of 'x' position variable %
%%%%%%%%%%%%%%%%%%%%%%%%%%%%%%%%%%%%%%%%%%%
t=[0:0.01:120]; a=1; w=pi;
yddot=a*w*cos(w*t);
y2ddot=-a*w*w*sin(w*t); yd=a*sin(w*t);
den=(sqrt(xddot.*xddot+yddot.*yddot)).*(sqrt(xddot.*xddo
t+yddot.*yddot)).*(sqrt(xddot.*xddot+yddot.*yddot));
     [t,y1]=ode45('yp1',[t],-5);
     [t,y2]=ode45('yp2',[t],-2);
     [t,y3]=ode45('yp3',[t],-7);
     [t,y4]=ode45('yp4',[t],-5);
% %% response for the 'x' variable %%%
x=y1;
y=y1.*y2-0.5;
z=y1.*y3-0.1;
Figure (1)
plot(t, x), grid, title(' Stabilization of 'x' variable')
%%%%%%%%%%%%%%%%%%%%%%%%%%%%%%%%%%%%%%%%%
% Stabilization of 'y' position variable %
%%%%%%%%%%%%%%%%%%%%%%%%%%%%%%%%%%%%%%%%%
t=[0:0.01:120]; a=1; w=pi;
x=t;
xddot=1;
yd=a*sin(w*t);
zd=1*ones(1,1001);
yddot=a*w*cos(w*t);
y2ddot=-a*w*w*sin(w*t); yd=a*sin(w*t);
den=(sqrt(xddot.*xddot+yddot.*yddot)).*(sqrt(xddot.*xddo
t+yddot.*yddot)).*(sqrt(xddot.*xddot+yddot.*yddot));
     [t,y1]=ode45('yp1',[t],-5);
     [t,y2]=ode45('yp2',[t],-2);
     [t,y3]=ode45('yp3',[t],-7);
     [t,y4]=ode45('yp4',[t],-5);
```

```
% %% response for the 'y' variable %%%
x=y1;
y=y1.*y2-0.5;
z=y1.*y3-0.1;
Figure (2)
plot(t, y), grid, title(' Stabilization of 'y' variable')

%%%%%%%%%%%%%%%%%%%%%%%%%
% Stabilization of 'z' position variable %
%%%%%%%%%%%%%%%%%%%%%%%%%
t=[0:0.01:120]; a=1; w=pi;
x=t;
xddot=1;
yd=a*sin(w*t);
zd=1*ones(1,1001);
yddot=a*w*cos(w*t);
y2ddot=-a*w*w*sin(w*t); yd=a*sin(w*t);
den=(sqrt(xddot.*xddot+yddot.*yddot)).*(sqrt(xddot.*xddo
t+yddot.*yddot)).*(sqrt(xddot.*xddot+yddot.*yddot));
     [t,y1]=ode45('yp1',[t],-5);
     [t,y2]=ode45('yp2',[t],-2);
     [t,y3]=ode45('yp3',[t],-7);
     [t,y4]=ode45('yp4',[t],-5);
% %% response for the 'z' variable %%%
x=y1;
y=y1.*y2-0.5;
z=y1.*y3-0.1;
Figure (3)
plot(t, z), grid, title(' Stabilization of 'z'
variable')
%%%%%%%%%%%%%%%%%%%%%%%%%
% Stabilization of 'ψ' position variable %
%%%%%%%%%%%%%%%%%%%%%%%%%
t=[0:0.01:120]; a=1; w=pi;
xd=t;
yd=a*sin(w*t);
zd=1;
xd2=a*w*cos(w*t);
xd4=0*ones(1,10000);
PSId=atan(xd2);
thetad=-atan(xd4'./sec(PSId));
phid=acot(cot(thetad)./sin(PSId)+tan(PSId)./sin(thetad)) ;
xddot=1;
yddot=a*w*cos(w*t);
y2ddot=-a*w*w*sin(w*t); yd=a*sin(w*t);
```

```
den=(sqrt(xddot.*xddot+yddot.*yddot)).*(sqrt(xddot.*xddo
t+yddot.*yddot)).*(sqrt(xddot.*xddot+yddot.*yddot));
      [t,y1]=ode45('yp1',[t],-5);
      [t,y2]=ode45('yp2',[t],-2);
      [t,y3]=ode45('yp3',[t],-7);
      [t,y4]=ode45('yp4',[t],-5);
x1=y1;
x2=y2;
x3=y1.*y2-0.5;
x4=y3;
x5=y1.*y3-0.1;
x6=y4;
% %% response for the 'PSI' variable %%%
PSI=atan(x2);
Figure (4)
plot(t, PSI), grid, title('Stabilization of 'psi'
variable')
%%%%%%%%%%%%%%%%%%%%%%%%%
% Stabilization of 'θ' position variable %
%%%%%%%%%%%%%%%%%%%%%%%%%
t=[0:0.01:120]; a=1; w=pi;
xd=t;
yd=a*sin(w*t);
zd=1;
xd2=a*w*cos(w*t);
xd4=0*ones(1,10000);
PSId=atan(xd2);
thetad=-atan(xd4'./sec(PSId));
phid=acot(cot(thetad)./sin(PSId)+tan(PSId)./sin(thetad)) ;
xddot=1;
yddot=a*w*cos(w*t);
y2ddot=-a*w*w*sin(w*t); yd=a*sin(w*t);
den=(sqrt(xddot.*xddot+yddot.*yddot)).*(sqrt(xddot.*xddo
t+yddot.*yddot)).*(sqrt(xddot.*xddot+yddot.*yddot));
      [t,y1]=ode45('yp1',[t],-5);
      [t,y2]=ode45('yp2',[t],-2);
      [t,y3]=ode45('yp3',[t],-7);
      [t,y4]=ode45('yp4',[t],-5);
x1=y1;
x2=y2;
x3=y1.*y2-0.5;
x4=y3;
x5=y1.*y3-0.1;
x6=y4;
% %% response for the 'Theta' variable %%%
PSI=atan(x2);
```

```
theta=-atan(x4./sec(PSI));
Figure (5)
plot(t, theta), grid, title('Stabilization of 'theta'
variable')
%%%%%%%%%%%%%%%%%%%%%%%%%
% Stabilization of 'ϕ' position variable %
%%%%%%%%%%%%%%%%%%%%%%%%%
t=[0:0.01:120]; a=1; w=pi;
xd=t;
yd=a*sin(w*t);
zd=1;
xd2=a*w*cos(w*t);
xd4=0*ones(1,10000);
PSId=atan(xd2);
thetad=-atan(xd4'./sec(PSId));
phid=acot(cot(thetad)./sin(PSId)+tan(PSId)./sin(thetad)) ;
xddot=1;
yddot=a*w*cos(w*t);
y2ddot=-a*w*w*sin(w*t); yd=a*sin(w*t);
den=(sqrt(xddot.*xddot+yddot.*yddot)).*(sqrt(xddot.*xddo
t+yddot.*yddot)).*(sqrt(xddot.*xddot+yddot.*yddot));
     [t,y1]=ode45('yp1',[t],-5);
     [t,y2]=ode45('yp2',[t],-2);
     [t,y3]=ode45('yp3',[t],-7);
     [t,y4]=ode45('yp4',[t],-5);
x1=y1;
x2=y2;
x3=y1.*y2-0.5;
x4=y3;
x5=y1.*y3-0.1;
x6=y4;
% %% response for the 'phi' variable %%%
PSI=atan(x2);
theta=-atan(x4./sec(PSI));
phi=acot(cot(theta)./sin(PSI)+tan(PSI)./sin(theta));
Figure (6)
plot(t, phi), grid, title('Stabilization of 'phi'
variable')
%%%%%%%%%%%%%%%%%%%%%%%%
% Stabilization of actual input v1%
%%%%%%%%%%%%%%%%%%%%%%%%
t=[0:0.01:120]; a=1; w=pi;
xd=t;
zd=1;
xd2=a*w*cos(w*t);
xd4=0*ones(1,10000);
```

```
PSId=atan(xd2);
thetad=-atan(xd4'./sec(PSId));
phid=acot(cot(thetad)./sin(PSId)+tan(PSId)./sin(thetad)) ;
xddot=1;
yd=a*sin(w*t);
yddot=a*w*cos(w*t);
y2ddot=-a*w*w*sin(w*t); yd=a*sin(w*t);
den=(sqrt(xddot.*xddot+yddot.*yddot)).*(sqrt(xddot.*xddo
t+yddot.*yddot)).*(sqrt(xddot.*xddot+yddot.*yddot));
     [t,y1]=ode45('yp1',[t],-5);
     [t,y2]=ode45('yp2',[t],-2);
     [t,y3]=ode45('yp3',[t],-7);
     [t,y4]=ode45('yp4',[t],-5);
x1=y1;
x2=y2;
x3=y1.*y2-0.5;
x4=y3;
x5=y1.*y3-0.1;
x6=y4;
PSI=atan(x2);
theta=-atan(x4./sec(PSI));
%%% Actual control variable v1 %%%
u1=-y1+0.26*(cos(t)-sin(t));
u2=-y2+2*0.5*(cos(t));
u3=-y3+2*0.1*(cos(t));
u4=-y4;
v1=u1./cos(PSI).*cos(theta);
Figure (7);
plot(t,v1), grid, title( 'original system input1 (v1));
%%%%%%%%%%%%%%%%%%%%%%%%%%%
% Stabilization of input v2 %
%%%%%%%%%%%%%%%%%%%%%%%%%%%%
t=[0:0.01:120]; a=1; w=pi;
xd=t;
xddot=1;
yd=a*sin(w*t);
zd=1*ones(1,1001);
yddot=a*w*cos(w*t);
y2ddot=-a*w*w*sin(w*t); yd=a*sin(w*t);
den=(sqrt(xddot.*xddot+yddot.*yddot)).*(sqrt(xddot.*xddo
t+yddot.*yddot)).*(sqrt(xddot.*xddot+yddot.*yddot));
     [t,y1]=ode45('yp1',[t],-5);
     [t,y2]=ode45('yp2',[t],-2);
     [t,y3]=ode45('yp3',[t],-7);
     [t,y4]=ode45('yp4',[t],-5);
x1=y1;
```

```
x2=y2;
x3=y1.*y2-0.5;
x4=y3;
x5=y1.*y3-0.1;
x6=y4;
PSI=atan(x2);
theta=-atan(x4./sec(PSI));
u1=-y1+0.26*(cos(t)-sin(t));
u2=-y2+2*0.5*(cos(t));
u3=-y3+2*0.1*(cos(t));
u4=-y4;
%%% Actual control variable v2 %%%
r11=cos(PSI).*cos(theta);
r23=sin(PSI).*sin(theta).*cos(phi)-cos(PSI).*sin(phi);
r33=cos(theta).*cos(phi);
r22=sin(PSI).*sin(theta).*sin(phi)+cos(PSI).*cos(phi);
r32=sin(phi).*cos(theta);
r12=cos(PSI).*sin(theta).*sin(phi)-sin(PSI).*cos(phi);
r13=sin(PSI).*sin(theta).*cos(phi)+sin(PSI).*sin(phi);
R=r11+r22+r33;
R1=1+R;
r11n=1./(1+r11);
v3=r11.*(-r23.*u2-r33.*u3);
v4=r11.*(r22.*u2+r32.*u3);
v2=r11n.*(R1.*u4-r12.*v3-r13.*v4);
Figure (8);
plot (t,v2),grid, title( 'original system input2 (v2');
%%%%%%%%%%%%%%%%%%%%%%%%%%%
% Stabilization of input v3 %
%%%%%%%%%%%%%%%%%%%%%%%%%%%
t=[0:0.01:120]; a=1; w=pi;
xd=t;
yd=a*sin(w*t);
xddot=1;
zd=1*ones(1,1001);
yddot=a*w*cos(w*t);
y2ddot=-a*w*w*sin(w*t); yd=a*sin(w*t);
den=(sqrt(xddot.*xddot+yddot.*yddot)).*(sqrt(xddot.*xddo
t+yddot.*yddot)).*(sqrt(xddot.*xddot+yddot.*yddot));
     [t,y1]=ode45('yp1',[t],-5);
     [t,y2]=ode45('yp2',[t],-2);
     [t,y3]=ode45('yp3',[t],-7);
     [t,y4]=ode45('yp4',[t],-5);
x1=y1;
x2=y2;
x3=y1.*y2-0.5;
```

```
x4=y3;
x5=y1.*y3-0.1;
x6=y4;
PSI=atan(x2);
theta=-atan(x4./sec(PSI));
u1=-y1+0.26*(cos(t)-sin(t));
u2=-y2+2*0.5*(cos(t));
u3=-y3+2*0.1*(cos(t));
u4=-y4;
%%% Actual control variable v3 %%%
r11=cos(PSI).*cos(theta);
r23=sin(PSI).*sin(theta).*cos(phi)-cos(PSI).*sin(phi);
r33=cos(theta).*cos(phi);
r22=sin(PSI).*sin(theta).*sin(phi)+cos(PSI).*cos(phi);
r32=sin(phi).*cos(theta);
r12=cos(PSI).*sin(theta).*sin(phi)-sin(PSI).*cos(phi);
r13=sin(PSI).*sin(theta).*cos(phi)+sin(PSI).*sin(phi);
R=r11+r22+r33;
R1=1+R;
r11n=1./(1+r11);
v3=r11.*(-r23.*u2-r33.*u3);
Figure (9);
plot (t,v3),grid, title( 'original system input2 (v3)');
%%%%%%%%%%%%%%%%%%%%%%%%%
% Stabilization of input v4 %
%%%%%%%%%%%%%%%%%%%%%%%%%%
t=[0:0.01:120]; a=1; w=pi;
xddot=1;
yddot=a*w*cos(w*t);
y2ddot=-a*w*w*sin(w*t); yd=a*sin(w*t);
den=(sqrt(xddot.*xddot+yddot.*yddot)).*(sqrt(xddot.*xddo
t+yddot.*yddot)).*(sqrt(xddot.*xddot+yddot.*yddot));
     [t,y1]=ode45('yp1',[t],-5);
     [t,y2]=ode45('yp2',[t],-2);
     [t,y3]=ode45('yp3',[t],-7);
     [t,y4]=ode45('yp4',[t],-5);
x1=y1;
x2=y2;
x3=y1.*y2-0.5;
x4=y3;
x5=y1.*y3-0.1;
x6=y4;
PSI=atan(x2);
theta=-atan(x4./sec(PSI));
u1=-y1+0.26*(cos(t)-sin(t));
u2=-y2+2*0.5*(cos(t));
```

```
u3=-y3+2*0.1*(cos(t));
u4=-y4;
%%% Actual control variable v4 %%%
r11=cos(PSI).*cos(theta);
r23=sin(PSI).*sin(theta).*cos(phi)-cos(PSI).*sin(phi);
r33=cos(theta).*cos(phi);
r22=sin(PSI).*sin(theta).*sin(phi)+cos(PSI).*cos(phi);
r32=sin(phi).*cos(theta);
r12=cos(PSI).*sin(theta).*sin(phi)-sin(PSI).*cos(phi);
r13=sin(PSI).*sin(theta).*cos(phi)+sin(PSI).*sin(phi);
R=r11+r22+r33;
R1=1+R;
r11n=1./(1+r11);
v4=r11.*(r22.*u2+r32.*u3);
Figure (10);
plot (t,v4),grid, title( 'original system input2 (v4)');
```

5 Control Design Using the Dynamic Model

In this chapter an overview of the dynamic mathematical model of an underwater vehicle is presented. The dynamic model is presented for the motion planning tasks. For the purpose of control design, the power form of the system is utilized.

5.1 DYNAMIC MODELING

In this section the dynamic model of the underwater vehicle is briefly discussed. The three-dimensional equations of motion for hydrodynamically shaped underwater vehicles can be developed using a body fixed coordinate frame and a global reference frame. The body fixed frame has components of motion given by the six velocity components $v = [v_x, v_y, v_z, \omega_x, \omega_y, \omega_z]$, relative to a constant velocity coordinate frame moving with the ocean current, v_c. The relative velocity is given as $v_r = v - v_c(q)$, and for a constant ocean current $v_r = v$ and the velocity vector is represented as

$$v = [v_x, v_y, v_z, \omega_x, \omega_y, \omega_z]^T \tag{5.1}$$

while the six components of position in the global reference frame are

$$q = [p \quad \eta]^T = [x, y, z, \phi, \theta, \psi]^T \tag{5.2}$$

The vehicle motion may be described in terms of the twelve nonlinear system equations [27] as

$$\begin{aligned} M(t)\dot{v} &= f(v(t), q(t), c(t)) + g(v(t), q(t)u_c(t) \\ \dot{q} &= h(v(t), q(t), v_c) \end{aligned} \tag{5.3}$$

The matrix $M(t)$ is a coupled mass matrix that includes both mechanical and hydrodynamic added mass; the functions f and g, which are mappings of the vehicle motions into forces, including Coriolis, gravitational, and centrifugal forces; the hydrostatic and hydrodynamic forces and moments acting on the vehicle in the body fixed coordinate frame, with coefficient c; the motion-dependent influence of control surfaces, thrusters, and any ballasting; and the function h, which includes the kinematical relationships found in performing the coordinate transformations between body fixed and global reference frames and the constant ocean current, v_c. The vector $u_c(t)$ is the control input vector from control surfaces, propeller speeds, thruster forces, and buoyancy adjustment in general. The equations of motion can be written in the simplified form as

$$\begin{aligned} M\dot{v} + C(v)v + D(v)v + g(q) &= \omega(\phi) + b(v,u) \\ \dot{q} &= G(q)v \end{aligned} \tag{5.4}$$

Here M is the inertial mass, including hydrodynamic virtual inertia or added mass; $C(v)$ contains the nonlinear forces and moments due to centrifugal and Coriolis forces; $D(v)$ is the vehicle's damping matrix, where the potential damping and viscous effect are lumped together; $g(q)$ is the vector containing the restoring terms formed by the vehicle's buoyancy and the gravitational terms; $\omega(\phi)$ is the wave and current disturbance vector; and $b(v,u)$ is a vector containing the vehicle's propulsion and control forces and moments. For this chapter the disturbance in the system is ignored. In Chapter 6 the effect of disturbance is incorporated back while studying the robust control. Ignoring the effect of disturbance, the dynamic model is given as

$$\begin{aligned} M\dot{v} + C(v)v + C_A(v)v + D(v)v + g(q) &= u_c \\ \dot{q} &= G(q)v \end{aligned} \tag{5.5}$$

where the control vector is defined as $u_c = b(v,u)$ and $\dot{q} = G(q)v$ is the kinematic model derived in Chapter 3, given by Equation 3.13. The matrix $G(q)$ is derived in Equation 3.12. Without any loss of generality, Equation 5.4 can be simplified as

$$\begin{aligned} M\dot{v} &= f(v,q) + u_c \\ \dot{q} &= G(q)v \end{aligned} \tag{5.6}$$

5.2 POINT-TO-POINT STABILIZATION CONTROL DESIGN

In the following section the control design for the problem of motion (point-to-point) stabilization is addressed. The system is supposed to reach a final desired configuration starting from an initial point, without the need to plan a trajectory. The control design task is to ensure the global asymptotic stability of the vectors $q(t)$ and $v(t)$. In this section we design a feedback control for the dynamic model of the underwater vehicles given by Equation 5.6 using the backstepping approach. The Lyapunov function will be used as a stability analysis tool as well as a feedback control design tool for this system. The design of feedback control is done in such a way that the Lyapunov function or its derivative has certain properties that guarantee boundedness or convergence to an equilibrium point.

5.2.1 State Feedback Control Using Backstepping

Here we address the problem of designing a state feedback controller $u_c(t)$ that stabilizes the origin $(q(t)=0, v(t)=0)$ of the control system (Equation 5.6). As stated in Chapter 4, the point stabilization for the kinematic model $\dot{q}=G(q)v$ cannot be achieved by a smooth time-invariant feedback. Only nonsmooth or time-varying feedback laws are of interest for the task. For our system we adopted the latter approach, where the time-varying smooth controller (Equation 4.39) was designed. The control law was designed utilizing the power form of the kinematic model. The power form as derived in Equation 4.38 is given as

$$
\begin{aligned}
\dot{y}_1 &= u_1 \\
\dot{y}_2 &= u_2 \\
\dot{y}_3 &= u_3 \\
\dot{y}_4 &= u_4;\ \dot{z}_{40}^{0} = u_4 \\
\dot{z}_{20}^{1} &= y_1 u_2 \\
\dot{z}_{20}^{1} &= y_1 u_2 \\
\dot{z}_{30}^{1} &= y_1 u_3
\end{aligned}
\tag{5.7}
$$

with the power form variables given as

$$
\begin{aligned}
y_1 &= x_1 \\
y_2 &= x_2 \\
y_3 &= x_4 \\
y_4 &= x_6;\ z_3 = z_{40}^{0} = x_6 \\
z_1 &= z_{20}^{1} = -x_3 + x_1 x_2 \\
z_2 &= z_{30}^{1} = -x_5 + x_1 x_4
\end{aligned}
\tag{5.8}
$$

Here vector $x=[x_1,x_2,x_3,x_4,x_5,x_6]^T$ is the chained-form variables given by the coordinate transformation in Equation 3.23 as

$$
\begin{aligned}
x_1 &= \xi_0 = x \\
x_2 &= \alpha_0 = \tan\psi \\
x_3 &= \alpha_1 = y \\
x_4 &= \eta_0 = -\tan\theta\sec\psi \\
x_5 &= z \\
x_6 &= \frac{1}{1+tr(R)}(r_{32}-r_{23})
\end{aligned}
\tag{5.9}
$$

The vector $u = [u_1, u_2, u_3, u_4]$ is the transformed input vector given in terms of the original velocity components $v = [v_x, v_y, v_z, \omega_x, \omega_y, \omega_z]$ by the input transformation (Equation 3.25) as

$$u_1 = \cos\psi\cos\theta v_1$$

$$u_2 = \sec^2\psi\sin\phi\sec c\theta v_3 + \sec^2\psi\cos\phi\sec\theta v_4$$

$$u_3 = \frac{(-\sin\psi\sin\phi\sin\theta - \cos\psi\cos\phi)}{\cos^2\psi\cos^2\theta}v_3 - \frac{(\sin\psi\cos\phi\sin\theta + \cos\psi\sin\phi)}{\cos^2\psi\cos^2\theta}v_4$$

$$u_4 = \frac{(1+\cos\psi\cos\theta)v_2 + (\cos\psi\sin\theta\sin\phi - \sin\psi\cos\phi)v_3 + (\sin\psi\sin\theta\cos\phi + \sin\psi\sin\phi)v_4}{1+\cos\psi\cos\theta + \sin\psi\sin\theta\sin\phi + \cos\psi\cos\phi + \cos\theta\cos\phi} \tag{5.10}$$

The power form (Equation 5.7) of the kinematic model $\dot{q} = G(q)v$ can be written in the simplified form as

$$\dot{y} = g(y)u \tag{5.11}$$

with vector $y = [y_1, y_2, y_3, y_4, z_1, z_2]^T$ denoting the power form variables and vector $u = [u_1, u_2, u_3, u_4]^T$ denoting the transformed velocity input vector. Using the power form of the kinematic model and the dynamic model, Equation 5.6 can be written as

$$\begin{aligned} M\dot{u} &= f(u,q) + u_c = \bar{u} \\ \dot{y} &= g(y)u \end{aligned} \tag{5.12}$$

where $\bar{u} = f(v,q) + u_c$ is the control variable. As stated earlier, the problem of motion stabilization for Equation 5.12 is to design a state feedback controller $\bar{u}(t)$ that stabilizes the origin $(y(t) = 0, u(t) = 0)$ of the control system (Equation 5.12). More specifically we consider the control law

$$\bar{u}(t) = F(y,u) \tag{5.13}$$

such that the origin of closed-loop dynamics is exponentially stable, with F being a nonlinear operator. The control strategy adopted here is similar in principle to feedback control by backstepping for ordinary differential equations [26]. Backstepping is a Lyapunov-based control method of feedback linearization. It is a recursive method that designs the feedback control law based on the choice of the Lyapunov function. It breaks the design problem for a system of equations into a sequence of design problems for scalar systems. We proceed with the control design as follows:

5.2.2 Control with Smooth Time-Varying Feedback

First we proceed to design a conceptual control law $u = H(y,t)$ for Equation 5.11 to stabilize the origin $y(t) = 0$, with $H(y,t)$ being a nonlinear map. Here $u(t)$ is an input to the power form of the kinematic model. This smooth time-varying control law was designed in Chapter 4 and is given by Equation 4.40 as

$$\begin{aligned} u_1 &= -y_1 + \sigma(\rho(z))(\cos t - \sin t) \\ u_2 &= -y_2 + c_1\sigma(z_1)\cos t \\ u_3 &= -y_3 + c_2\sigma(z_2)\cos t \\ u_4 &= -y_4 \end{aligned} \tag{5.14}$$

where $c_j > 0$ and $\sigma : R \to R$ is a nondecreasing C^3 saturation function with a magnitude less than some $\delta > 0$ and is linear between $(-\delta, \delta)$. For global stabilization δ should be small enough. The saturation function then satisfies the following:

1. $\sigma(z) = z$ when $|z| \leq \varepsilon$
2. $|\sigma(z)| \leq \delta$ for all z such that $0 < \varepsilon < \delta$

With the controller (Equation 5.14) we get the conceptual closed-loop dynamics for Equation 5.11 as

$$\begin{aligned} \dot{y}_1 &= -y_1 + \sigma(\rho(z))(\cos t - \sin t) \\ \dot{y}_2 &= -y_2 + c_1\sigma(z_1)\cos t \\ \dot{y}_3 &= -y_3 + c_2\sigma(z_2)\cos t \\ \dot{y}_4 &= -y_4 \\ \dot{z}_1 &= -y_1 y_2 + y_1 c_1\sigma(z_1)\cos t \\ \dot{z}_2 &= -y_1 y_3 + y_1 c_2\sigma(z_2)\cos t \end{aligned} \tag{5.15}$$

5.2.3 Lyapunov Stability Analysis

In order to check the stability of the closed-loop system (Equation 5.15), we use the Lyapunov function analysis method. The stability problem is to establish sufficient conditions for which the origin $y(t) = 0$ of the closed-loop dynamics (Equation 5.15) is globally asymptotic stable. Within the framework of our system the definition of Lyapunov stability can be given as follows:

Definition: An equilibrium state $y = 0$ of a dynamical system (Equation 5.15) is stable if for each real number $\varepsilon > 0$, there exists a real number $\delta = \delta(\varepsilon) > 0$ such that

$$\|y(t_0)\| < \delta \Rightarrow \|y(t)\| < \varepsilon \quad \forall t \geq t_0 \tag{5.16}$$

If in addition δ can be chosen such that

$$\|y(t_0)\| < \delta \Rightarrow \lim_{t\to\infty} y(t) = 0 \tag{5.17}$$

then the equilibrium is said to be asymptotically stable.

The stability of an equilibrium point $y = 0$ of a dynamical system (Equation 5.15) can also be evaluated by defining a Lyapunov function $V(t)$ for this system. Here $V : D \to \Re$ is a continuously differentiable function defined in a domain $D \subset \Re^6$ that contains the origin. In terms of the Lyapunov function the stability can be defined as follows:

Definition: An equilibrium state $y = 0$ of a dynamical system (Equation 5.15) is stable if there exists a continuously differentiable function $V : D \to \Re$ such that

$$V(0) = 0 \quad \text{and} \quad V(y) > 0 \quad \forall y \neq 0 \tag{5.18}$$

$$\dot{V}(y) \leq 0 \quad \forall y \in D \tag{5.19}$$

If in addition

$$\dot{V}(y) < 0 \quad \forall y \neq 0 \tag{5.20}$$

then the equilibrium is said to be asymptotically stable.

Let us consider a Lyapunov function $V : D \to \Re$ for the system in Equation 5.15 as

$$V(y,t) = \frac{1}{2} y^T(t) y(t) \tag{5.21}$$

Since $V(y)$ is a quadratic function of $V(t)$, we can easily conclude that $V(y)$ is a positive definite function. The time rate of change of $V(y)$ is given as

$$\dot{V}(y,t) = \frac{\partial V(y)}{\partial y} \dot{y}(t) = \sum_{i=1}^{6} \frac{\partial V(y)}{\partial y_i} \dot{y}_i \tag{5.22}$$

Using the center manifold theorem, it can be proven that the derivative of the Lyapunov function $\dot{V}(y)$ given by Equation 5.22 for the closed-loop dynamics (Equation 5.15) is a negative definite function. Hence, it can be concluded that the equilibrium point of dynamics given by Equation 5.15, using the control law in Equation 5.14, is asymptotically stable. Also, for some $0 < \varepsilon < \delta$, $\exists \delta_0$ such that if $\varepsilon < \varepsilon_0$, the closed-loop dynamics are globally asymptotically stabilized to zero and $\dot{V}(y)$ is bounded by a negative definite function $-W(y)$. The detailed proof is given in [28].

5.2.4 Control of the Dynamic Model

In the previous section we established the asymptotic stability of the origin $y(t) = 0$ for the kinematic model

$$\dot{y} = g(y)u \tag{5.23}$$

using the conceptual control law

$$u(t) = H(y,t) \tag{5.24}$$

so that the closed-loop dynamics are given by Equation 5.15 as

$$\dot{y} = g(y)H(y,t) \tag{5.25}$$

As already shown, this closed-loop system is asymptotically stable with a conceptual Lyapunov function (Equation 5.21)

$$V(y,t) = \frac{1}{2} y^T(t) y(t)$$

We have used the term *conceptual* with the control law, closed-loop system, and Lyapunov function. This is done in order to stress the fact that the control law in Equation 5.24 cannot be implemented in practice, as $u(t)$ is not a control variable. However, this conceptual design helps us to recognize the benefit of the input $u(t)$ being close to $H(y,t)$. From the knowledge of the conceptual Lyapunov function $V(t)$ we want to design a smooth feedback control for stabilizing the origin of the overall system (Equation 5.12). We therefore add to the conceptual Lyapunov function (Equation 5.21) a term penalizing the difference between $u(t)$ and $H(y,t)$. For this purpose, we rewrite the dynamics in Equation 5.24 as

$$\dot{y} = g(y)H(y,t) + g(y)\{u(t) - H(y,t)\} \tag{5.26}$$

Defining the difference between $u(t)$ and $H(y,t)$ by an error variable $l(t)$ $u(t) - H(y,t)$, we get the following modified dynamics:

$$\dot{y}(t) = g(y)H(y,t) + g(y)l(t) \tag{5.27}$$

$$\dot{l}(t) = \dot{u}(t) - \dot{H}(y,t) \tag{5.28}$$

Substituting the dynamic model in Equation 5.12 in Equation 5.28, we get the following dynamic model:

$$\dot{y}(t) = g(y)H(y,t) + g(y)l(t) \tag{5.29}$$

$$\dot{l}(t) = \frac{\bar{u}(t)}{M} - \dot{H}(y,t) = u_n(t) \tag{5.30}$$

where $u_n(t) = \frac{\bar{u}(t)}{M} - \dot{H}(y,t)$ is the new control variable. The above system is similar to the system in Equation 5.12, except that the first component (Equation 5.29) has an asymptotically stable origin when its input $l(t)$ is zero. The time derivative of function $H(y,t)$ can be computed using the following expression:

$$\dot{H}(y,t) = \frac{\partial H}{dy}\dot{y}(t) = \frac{\partial H}{dy} g(y)u \tag{5.31}$$

Now let us modify the Lyapunov function (Equation 5.21) by adding an error term to it, thus resulting in the Lyapunov function for the overall system as

$$\begin{aligned} V_a(y,u) &= V(y)+\frac{1}{2}l^T(t)l(t) \\ &= \frac{1}{2}y^T(t)y(t)+\frac{1}{2}l^T(t)l(t) \end{aligned} \tag{5.32}$$

The time rate of change of this functional using Equations 5.29 and 5.30 is given as

$$\begin{aligned} \dot{V}_a(y,u) &= \frac{\partial V}{\partial y}\dot{y}(t)+l^T(t)\dot{l}(t) \\ &= \frac{\partial V}{\partial y}[g(y)H(y,t)]+\frac{\partial V}{\partial y}[g(y)l(t)]+l^T(t)u_n(t) \end{aligned} \tag{5.33}$$

The first term in the above equation is bounded by a negative definite function $-W(y)$; therefore, we have

$$\dot{V}_a(y,u) \leq -W(y)+\frac{\partial V}{\partial y}[g(y)l(t)]+l^T(t)u_n(t) \tag{5.34}$$

We have to choose a new control law $u_n(t)$ in such a manner that the time derivative of the new function or the sum of the second and third terms is also bounded by a negative definite function. By choosing the control law

$$u_n(t) = -\frac{\partial V}{\partial y}g(y)-kl(t), \quad k>0 \tag{5.35}$$

we get the following result:

$$\dot{V}_a(y,u) \leq -W(y)-kl^T(t)l(t) \tag{5.36}$$

Since the product $l^T(t)l(t)$ is positive definite, with $k > 0$, the origin ($y(t) = 0$, $l(t) = 0$) of the system in Equations 5.29 and 5.30 is asymptotically stable. The feedback control law is given by

$$\begin{aligned} u_n(t) &= -\frac{\partial V}{\partial y}g(y)-kl(t) \\ &= \sum_{i=1}^{6}\frac{\partial V(y)}{\partial y_i}g(y)-kl(t) \\ &= \sum_{i=1}^{6}y_i g(y)-kl(t) \end{aligned} \tag{5.37}$$

where $g(y)$ is the nonlinear system function for Equation 5.7 and $l(t) = u(t) - H(y,t)$, with $u(t)$being the control law given by Equation 5.14. The feedback controller for the original system (Equation 5.12) is given by

$$\frac{\bar{u}(t)}{M} = u_n(t) + \dot{H}(y,t)$$
$$\bar{u}(t) = M[u_n(t) + \dot{H}(y,t)] \tag{5.38}$$

Using Equation 5.31,

$$\dot{H}(y,t) = \frac{\partial H}{dy}\dot{y}(t) = \frac{\partial H}{dy}g(y)u$$

the control law $\bar{u}(t)$ can be computed as

$$\bar{u}(t) = M[u_n(t) + \frac{\partial H}{dy}g(y)u(t)] \tag{5.39}$$

with the control $u_n(t)$ given by Equation 5.37 and the control law $u(t)$ given by Equation 5.14. With the feedback controller (Equation 5.37) we have proved that the origin $(y(t) = 0, l(t) = 0)$ of Equations 5.29 and 5.30 is asymptotically stable. But we need to prove the asymptotic stability of the origin $(y(t) = 0, u(t) = 0)$ of the system (Equation 5.12), or we need to prove the asymptotic stability of $u(t) = 0$ using the control law in Equation 5.39. The proposition below establishes the asymptotic stability of the original closed-loop system (Equation 5.12).

Proposition: The origin $(y(t) = 0, u(t) = 0)$ of the system in Equation 5.12 is asymptotically stable under the control law $\bar{u}(t) = M[u_n(t) + \frac{\partial H}{dy}g(y)u(t)]$, with $u_n(t)$ given by Equation 5.37 and control law $u(t)$ given by Equation 5.14.

Proof: The control laws in Equation 5.37 make the origin $(y(t) = 0, l(t) = 0$ for Equations 5.29 and 5.30 asymptotically go to zero. From the equation $l(t) = u(t) - H(y,t)$, we want to prove the asymptotic stability of $u(t) = 0$. We know that the control $H(y,t)$ is zero for $y(t) = 0$, that is, $H(0) = 0$. We can conclude that the origin $(y(t) = 0$ $u(t) = 0)$ is asymptotically stable using the control law in Equation 5.39.

6 Robust Feedback Control Design

In previous chapters we addressed the control of underwater vehicles without accounting for the presence of uncertainty in the design of the controller. The uncertainty is a mismatch between the model used for controller design and the actual process model. The uncertain function may include uncertain model parameters or external disturbances. Here we consider the case where we have an uncertainty in the input to the system. The objective is to develop a framework for the synthesis of robust controllers that handle the effect of this uncertain variable for both kinematic and dynamic models of an underwater vehicle. The robust controllers designed guarantee boundedness of state and achieve asymptotic stabilization with an arbitrary degree of asymptotic attenuation to the effect of uncertain variables on the output of the closed-loop system. The controllers are designed constructively using Lyapunov's direct method [31] and require the existence of known bounding functions that capture the magnitude of the uncertain term. This chapter presents the design of robust nonlinear feedback controllers for models given in Chapters 3 and 5 representing kinematic and dynamic models for underwater vehicles, with uncertainty in the control input. In both cases the objective of control design is to synthesize nonlinear feedback controllers that stabilize the system and guarantee the stability of the closed-loop system in the presence of uncertainty.

6.1 ROBUST CONTROL USING THE KINEMATIC MODEL

In this section we formulate the uncertain control model and present robust control design for the kinematic model derived in Chapter 3. The problem is to design a state feedback control for the problem of motion (point-to-point) stabilization. The control design task is to ensure the global asymptotic stability of the vector $q(t)$ of the closed-loop system irrespective of uncertain elements. The kinematic model is given by Equation 3.13 as

$$\dot{q} = G(q)v \tag{6.1}$$

Here the vector $q = [p \quad \eta]^T = [x, y, z, \phi, \theta, \psi]^T$ is the six components of position in the global reference frame variable we want to control. The body fixed frame has components of motion given by the six-velocity-component vector $v = [v_x, v_y, v_z, \omega_x, \omega_y, \omega_z]$, relative to a constant velocity coordinate frame moving with the ocean current, v_c. As stated in Chapter 4, the point stabilization for the kinematic model $\dot{q} = G(q)v$ cannot be achieved by a smooth time-invariant feedback. Only nonsmooth or time-varying feedback laws are of interest for the task. For our system we adopted the latter approach, where the time-varying smooth controller (Equation 4.39) was designed. The control law was designed utilizing the power form of the kinematic model. The power form as derived in Equation 4.38 is given as

$$\begin{aligned} \dot{y}_1 &= u_1 \\ \dot{y}_2 &= u_2 \\ \dot{y}_3 &= u_3 \\ \dot{y}_4 &= u_4;\ \dot{z}_{40}^0 = u_4 \\ \dot{z}_{20}^1 &= y_1 u_2 \\ \dot{z}_{30}^1 &= y_1 u_3 \end{aligned} \tag{6.2}$$

with the power form variables given as

$$\begin{aligned} y_1 &= x_1 \\ y_2 &= x_2 \\ y_3 &= x_4 \\ y_4 &= x_6,\ z_3 = z_{40}^0 = x_6 \\ z_1 &= z_{20}^1 = -x_3 + x_1 x_2 \\ z_2 &= z_{30}^1 = -x_5 + x_1 x_4 \end{aligned} \tag{6.3}$$

Here vector $x = [x_1, x_2, x_3, x_4, x_5, x_6]^T$ is the chained-form variables given by the coordinate transformation in Equation 3.23. The vector $u = [u_1, u_2, u_3, u_4]$ is the transformed input vector given in terms of the original velocity components $v = [v_x, v_y, v_z, \omega_x, \omega_y, \omega_z]$ by the input transformation (Equation 3.25). The power form (Equation 6.2) of the kinematic model $\dot{q} = G(q)v$ can be written in the simplified form as

$$\dot{y} = g(y)u \tag{6.4}$$

with vector $y = [y_1, y_2, y_3, y_4, z_1, z_2]^T$ denoting the power form variables and vector $u = [u_1, u_2, u_3, u_4]^T$ denoting the transformed velocity input vector.

6.1.1 Input Uncertain Control Model

Let us consider that the uncertain variable for this system is the control input or the free flow velocity. This means that there is an error between the control command and the actual input command to the system. The uncertain model is therefore given as

$$\dot{y} = g(y)[u(t) + \theta] \tag{6.5}$$

In the above equation $\theta = \theta(t,u,y)$ denotes the unknown function, which takes care of uncertainty in the input $u(t)$ to the system. The uncertain term here satisfies an important structural property; namely, it enters the state equation exactly at the point where the control variable enters. This property will be referred to as the matching condition. The nominal model (the system without uncertainty) for this system is described by

$$\dot{y} = g(y)u$$

The first step is to design a stabilizing feedback controller for the nominal model. We proceed to design a conceptual control law $u(t) = H(y,t)$ for Equation 6.4 to stabilize the origin $y(t) = 0$, with $H(y,t)$ being a nonlinear map and $u(t)$ an input to the power form of the kinematic model. This smooth time-varying control law was designed in Chapter 4 and is given by Equation 4.40 as

$$\begin{aligned} u_1 &= -y_1 + \sigma(\rho(z))(\cos t - \sin t) \\ u_2 &= -y_2 + c_1\sigma(z_1)\cos t \\ u_3 &= -y_3 + c_2\sigma(z_2)\cos t \\ u_4 &= -y_4 \end{aligned} \tag{6.6}$$

where $c_j > 0$ and $\sigma: R \to R$ is a nondecreasing C^3 saturation function with a magnitude less than some $\delta > 0$ and is linear between $(-\delta,\delta)$. For global stabilization δ should be small enough. The saturation function then satisfies the following:

1. $\sigma(z) = z$ when $|z| \le \varepsilon$
2. $|\sigma(z)| \le \delta$ for all z such that $0 < \varepsilon < \delta$

With the controller (Equation 6.6) we get the nominal closed-loop dynamics for Equation 6.2 as

$$\begin{aligned} \dot{y}_1 &= -y_1 + \sigma(\rho(z))(\cos t - \sin t) \\ \dot{y}_2 &= -y_2 + c_1\sigma(z_1)\cos t \\ \dot{y}_3 &= -y_3 + c_2\sigma(z_2)\cos t \\ \dot{y}_4 &= -y_4 \\ \dot{z}_1 &= -y_1y_2 + y_1c_1\sigma(z_1)\cos t \\ \dot{z}_2 &= -y_1y_3 + y_1c_2\sigma(z_2)\cos t \end{aligned} \tag{6.7}$$

Thus, with $|\theta(t,u,y)| = 0$ and $u(t) = H(y,t)$ the nominal closed-loop system $\dot{y}(t) = g(y)$ $H(y,t)$ given by Equation 6.7 has a globally asymptotic stable origin, as proved in Section 5.2.3, and there exists a Lyapunov function given by Equation 5.21,

$$V(y,t) = \frac{1}{2}y^T(t)y(t)$$

whose time derivative $\dot{V}(y)$ is bounded by a negative definite function $-W(y)$ as

$$\dot{V}(y) \leq -W(y) \tag{6.8}$$

with $W(y)$ being a positive definite function.

6.1.2 Robust Control by the Lyapunov Redesign Method

In this section we consider the system of Equation 6.5 and address the problem of synthesizing a state feedback controller that stabilizes the closed-loop system irrespective of the uncertainty. The controller is designed constructively using Lyapunov's direct method, where we use the Lyapunov function for the system to design feedback control. A standard method for finding a Lyapunov function for an uncertain system is developed in [29] and is known as *Lyapunov redesign*. This technique has been incorporated in various books like [30], [31], and [32]. The key idea of this method is to employ a Lyapunov function for the nominal system as a Lyapunov function for the uncertain system. This reuse of the same Lyapunov function is referred to by the term *redesign*. The Lyapunov redesign technique uses a Lyapunov function of a nominal system to design an additional control component to robustify the design to a class of large uncertainties that satisfy the matching condition, that is, the uncertain terms enter the state equation at the same point as the input. Lyapunov redesign can be used to achieve robust stabilization. The goal is to design a feedback control law for Equation 6.5 as

$$u(t) = H(y,t) + G(y,t) \tag{6.9}$$

such that we achieve closed-loop stability and asymptotic attenuation of $\theta(t,u,y)$, where $G(y,t)$ is a nonlinear operator mapping. In Equation 6.9 $H(y,t)$ achieves closed-loop stability and $G(y,t)$ asymptotically attenuates the effect of $\theta(t,u,y)$. The function $H(y,t)$ will be designed by the previous approach, and $G(y,t)$ will be designed using the Lyapunov redesign method. The design of function $G(y,t)$ is known as Lyapunov redesign [26] and is done on the basis of the assumption that we have a bounding function that captures the size of the disturbance. Let us assume with the controller in Equation 6.9 that there exists a known smooth function that bounds the magnitude of uncertain variables as

$$\begin{aligned} \|\theta(t,u,y)\| &= \|\theta(t,y,(H(y,t)+G(y,t))\| \\ &\leq \gamma(t,y) + \kappa \,\|G(y,t))\| \end{aligned} \tag{6.10}$$

where $\gamma(t,y)$ is a nonnegative function and is a measure of size of uncertainty $\theta(t,u,y)$. From Equation 6.10, the only requirement is the knowledge of $\gamma(t,y)$, which doesn't necessarily have to be small. From the knowledge of the Lyapunov function $V(t)$ and functions $\gamma(t,y)$ and κ the goal is to design $G(y,t)$ and apply $u(t) = H(y,t) + G(y,t)$ to the actual system (Equation 6.5) such that the overall closed-loop system is stabilized in the presence of uncertainty. By using the control law in Equation 6.6 the feedback control law (Equation 6.9) is given by

$$u(t) = H(y,t) + G(y,t)$$

Under the feedback control law the closed-loop dynamics for Equation 6.5 take the form

$$\dot{y}(t) = g(y)H(y,t) + g(y)G(y,t) \tag{6.11}$$

Thus, the error in the nominal closed-loop dynamics $\dot{y}(t) = g(y)H(y,t)$ given by Equation 6.7 and the closed-loop dynamics (Equation 6.11) due to input uncertainty is given by

$$\theta = g(y)G(y,t) \tag{6.12}$$

with $g(y)$ given by Equation 6.2 as

$$g(y) = \begin{bmatrix} 1 & 0 & 0 & 0 & 0 & 0 \\ 0 & 1 & 0 & 0 & 0 & 0 \\ 0 & 0 & 1 & 0 & 0 & 0 \\ 0 & 0 & 0 & 0 & 1 & 0 \\ 0 & y_1 & 0 & 0 & 0 & 0 \\ 0 & 0 & y_1 & 0 & 0 & 0 \end{bmatrix} \tag{6.13}$$

Let us denote the component $G(y,t)$ of control input in Equation 6.9; by v we can rewrite Equation 6.12 as

$$\theta = g(y)v$$

The magnitude of this uncertain term is given by

$$\|\theta\| = \|g(y)v\| \tag{6.14}$$

The bound on the magnitude of uncertainty (Equation 6.14) can be found as

$$\begin{aligned} \|\theta(t,y,u)\| &\le \|g(y)\| \, \|v\| \\ &= \gamma(y,t) + \kappa(y,t)v \end{aligned} \tag{6.15}$$

where $\kappa(y,t) = y_1^2$, $\gamma(y,t) = \|g(y)\|$. This bound will be utilized to design the control law $G(y,t)$. Let us now apply the control law in Equation 6.9 to the input uncertain system (Equation 6.5) so that the closed-loop dynamics take the form

$$\dot{y}(t) = g(y)H(y,t) + g(y)[G(y,t) + \theta(t,y,u)] \tag{6.16}$$

As can be seen, the system in Equation 6.16 is a perturbation of the nominal closed-loop system (Equation 6.7), the third term being the perturbation. Let us choose the Lyapunov function $V(y,t) = \frac{1}{2}y^T(t)y(t)$ for Equation 6.16, same as before. To design $G(y,t)$ we find the rate of change of the Lyapunov function as

$$\dot{V}(y,t) = \frac{\partial V(y)}{\partial y}\dot{y}(t) = \sum_{i=1}^{6}\frac{\partial V(y)}{\partial y_i}\dot{y}_i$$

$$\begin{aligned}\dot{V}(y,t) &= \frac{\partial V(y)}{\partial y}\{g(y)H(y,t) + g(y)[G(y,t) + \theta(t,y,u)]\}\\ &= \frac{\partial V(y)}{\partial y}[g(y)H(y,t)] + \frac{\partial V(y)}{\partial y}\{g(y)[G(y,t) + \theta(t,y,u)]\}\\ &\le -W(y) + \frac{\partial V(y)}{\partial y}\{g(y)[G(y,t) + \theta(t,y,u)]\}\end{aligned} \tag{6.17}$$

The first inequality is because of the asymptotic stability of the nominal closed-loop system (Equation 6.7). We need to choose a control law $G(y,t)$ so as to cancel the destabilizing effort of $\theta(t,u,y)$ on $\dot{V}(y,t)$. The law $G(y,t)$ should be such that the second term in Equation 6.17 is negative semidefinite in order to have asymptotic stability. By denoting $G(y,t)$ by v we can rewrite Equation 6.17 as

$$\dot{V}(y,t) \le -W(y) + \frac{\partial V(y)}{\partial y}[g(y)v(t)] + \frac{\partial V(y)}{\partial y}[g(y)\theta(t)] \tag{6.18}$$

$$\dot{V}(y,t) \le -W(y) + \sum_{i=1}^{6}\frac{\partial V(y)}{\partial y_i}g(y)v(t) + \sum_{i=1}^{6}\frac{\partial V(y)}{\partial y_i}[g(y)\theta(t)] \tag{6.19}$$

By using Equation 6.15 we have the rate of change of Lyapunov bounded as

$$\begin{aligned}\dot{V}(y,t) &\le -W(y) + \sum_{i=1}^{6}\frac{\partial V(y)}{\partial y_i}g(y)v(t) + \sum_{i=1}^{6}\frac{\partial V(y)}{\partial y_i}[g(y)\{\gamma(y,t) + \kappa(y,t)v\}]\\ &\le -W(y) + \sum_{i=1}^{6}\frac{\partial V(y)}{\partial y_i}g(y)[1 + \kappa(y,t)]v(t) + \sum_{i=1}^{6}\frac{\partial V(y)}{\partial y_i}[g(y)\gamma(y,t)]\end{aligned} \tag{6.20}$$

Denoting

$$w^T(y) = \frac{\partial V(y)}{\partial y} g(y) = \sum_{i=1}^{6} \frac{\partial V(y)}{\partial y_i} g(y) \tag{6.21}$$

we can rewrite Equation 6.20 as

$$\dot{V}(y,t) \leq -W(y) + w^T(y)[1+\kappa(y,t)]v(t) + w^T(y)\gamma(y,t) \tag{6.22}$$

Now let us choose the state feedback control as follows:

$$v(t) = -\frac{\eta(y,t)}{1-\kappa(y,t)} w(y,t) \tag{6.23}$$

where $\eta(y,t) \geq \gamma(y,t)$ for all $t \geq 0$ is a nonnegative function. By using this control law it is seen that the sum of the second and third terms in Equation 6.22 is zero, that is,

$$w^T(y)[1+\kappa(y,t)]v(t) + w^T(y)\gamma(y,t) = 0 \tag{6.24}$$

By using the control law in Equation 6.23 and then applying Equation 6.24 we can show that

$$\dot{V}(y,t) \leq -W(y))$$

Hence, with the control law in Equation 6.23, the derivative of $V(y,t)$ for the closed-loop system (Equation 6.16) is negative definite and we have asymptotic stability in the presence of the disturbance. The robust control law for Equation 6.5 is therefore given as

$$u(t) = H(y,t) - \frac{\eta(y,t)}{1-\kappa(y,t)} w(y,t) \tag{6.25}$$

6.2 ROBUST CONTROL USING THE DYNAMIC MODEL

In this section we formulate the control model and present a robust control design for the dynamic model of an underwater vehicle. This section also studies Lyapunov stability for this model with the uncertainty in the input to the system. The control design is done using a combination of both *Lyapunov redesign* and *backstepping*, and the technique is referred to as *robust backstepping.* Let us consider the dynamic model for an underwater vehicle given in Section 5.21 by Equation 5.12 as

$$M\dot{u} = f(u,q) + u_c = \bar{u} \tag{6.26}$$

$$\dot{y} = g(y)u \tag{6.27}$$

where $\bar{u} = f(v,q) + u_c$ is the control variable. Here $y(t)$ and $u(t)$ are the variables we want to control. For point-to-point stabilization we require $y(t) = 0$ and $u(t) = 0$.

Now let us consider the uncertainty in the control input for this system. We can have both matched and unmatched uncertainties, depending upon which equation contains the uncertainty. We can have uncertainty in Equation 6.27 where it enters the equation through the control variable $u(t)$. For this case the uncertainty is unmatched since $u(t)$ is the conceptual control input to this equation and is not the actual control input as in Equation 6.26. We can also have uncertainty in the control variable $\bar{u}(t)$, in which case the uncertainty is matched and we can also have both. The control design scheme in all cases is that of robust backstepping [32], where we do a combination of backstepping and Lyapunov redesign. Now let us discuss these cases one by one.

6.2.1 Robust Backstepping: Unmatched Uncertainty

Let us first consider the uncertainty in the input $u(t)$ to the nominal system in Equation 6.27. The control design strategy here is robust backstepping, which is the combination of backstepping and Lyapunov redesign. The uncertain model is given as

$$M\dot{u} = \bar{u} \tag{6.28}$$

$$\dot{y} = g(y)[u(t) + \theta] \tag{6.29}$$

where $\theta = \theta\ (t,u,y)$ is the unknown function that takes care of the disturbance in the input $u(t)$ to Equation 6.27. The uncertain term for the dynamic system in Equation 6.28 is unmatched, as it does not enter the system at the same point as the input $\bar{u}$, even though it enters equation Equation 6.26 through the conceptual input $u(t)$. As a first step, we design a robust control for Equation 6.29 using the Lyapunov redesign technique, and in the next step find the control law for the overall system (Equation 6.28) using backstepping.

1. **Lyapunov redesign:** The first step is to design a robust controller for Equation 6.29 using the Lyapunov redesign method. The goal is to design a robust control law for Equation 6.30 as

$$u(t) = H(y,t) + v(t) = \bar{H}(y,t) \tag{6.30}$$

such that we achieve closed-loop stability and asymptotic attenuation of $\theta(t,u,y)$. $H(y,t)$ in Equation 6.30 achieves closed-loop stability and $v(t)$ asymptotically attenuates the effect of $\theta(t,u,y)$. The function $H(y,t)$ is designed from the feedback linearization of the nominal model in Equation 6.27 for

Equation 6.29, and $v(t)$ will be designed by using the Lyapunov redesign method. The nominal model (Equation 6.27) is given as

$$\dot{y} = g(y)u$$

The stabilizing feedback controller $u(t) = H(y,t)$ for this nominal model from Equation 4.40 given in Chapter 4 is

$$\begin{aligned}
u_1 &= -y_1 + \sigma(\rho(z))(\cos t - \sin t) \\
u_2 &= -y_2 + c_1\sigma(z_1)\cos t \\
u_3 &= -y_3 + c_2\sigma(z_2)\cos t \\
u_4 &= -y_4
\end{aligned} \tag{6.31}$$

where $c_j > 0$ and $\sigma : R \to R$ is a nondecreasing C^3 saturation function with a magnitude less than some $\delta > 0$ and is linear between $(-\delta,\delta)$. For global stabilization δ should be small enough. The saturation function then satisfies the following:

1. $\sigma(z) = z$ when $|z| \leq \varepsilon$
2. $|\sigma(z)| \leq \delta$ for all z such that $0 < \varepsilon < \delta$

which gives the conceptual nominal closed-loop system as given by Equation 6.7:

$$\dot{y}(t) = g(y)H(y,t) \tag{6.32}$$

The nominal closed-loop model (Equation 6.32) has an asymptotically stable origin, and there exists a Lyapunov function $V(t)$ given by Equation 5.21

$$V(y,t) = \frac{1}{2}y^T(t)y(t)$$

which satisfies

$$\dot{V}(y) \leq -W(y) \tag{6.33}$$

with $W(y)$ being a positive definite function. Now by using Equation 6.31 the control law (Equation 6.30) becomes

$$u(t) = H(y,t) + v(t) \tag{6.34}$$

which gives the following closed-loop dynamics for Equation 6.27:

$$\dot{y}(t) = g(y)H(y,t) + g(y)v(t) \tag{6.35}$$

Thus, the error in the closed-loop dynamics (Equations 6.32 and 6.35) due to input uncertainty is given by

$$\theta(t,y,u) = g(y)v(t) \tag{6.36}$$

with $g(y)$ given by Equation 6.13. The magnitude of this uncertain term due to Equation 6.15 is

$$\begin{aligned} \|\theta(t,y,u)\| &\leq \|g(y)\|\,\|v(t)\| \\ &= \gamma(y,t) + \kappa(y,t)\,\|v(t)\| \end{aligned} \tag{6.37}$$

From the previous subsection we found the control law $v(t)$ using Lyapunov redesign, given by Equation 6.23 as

$$v(t) = -\frac{\eta(y,t)}{1-\kappa(y,t)} w(y,t) \tag{6.38}$$

where $\eta(y,t) \geq \gamma(y,t)$ for all $t \geq 0$ is a nonnegative function. By using this control law, it is seen that the closed-loop system in Equation 6.29 has asymptotic stability in the presence of the disturbance. The robust control law for Equation 6.29 is therefore given as

$$u(t) = H(y,t) - \frac{\eta(y,t)}{1-\kappa(y,t)} w(y,t) \tag{6.39}$$

2. **Control by backstepping method:** The next step is to design a feedback controller for the overall system (Equation 6.28) using the backstepping method described in Section 5.2.1. The goal is to a design feedback control law $\bar{u}(t)$ to stabilize the overall system from the knowledge of the Lyapunov function $V(t)$ for Equation 6.29 and modify it. We proceed with the control design in a similar way as in Section 5.2.1, as follows:
   a. First design a conceptual control law $u(t) = \bar{H}(y,t)$ for Equation 6.29 from the previous step to stabilize origin $y(t) = 0$. With this control law the conceptual closed-loop dynamics for Equation 6.29 are asymptotically stable and there exists a Lyapunov function (Equation 5.21) that satisfies $\dot{V}(y,t) \leq -W(y)$.
   b. Since $u(t)$ is not the actual control variable, defining the difference between $u(t)$ and $\bar{H}(y,t)$ by an error variable $l(t) = u(t) - \bar{H}(y,t)$, we get the following modified dynamics:

$$\dot{y}(t) = g(y)[\bar{H}(y,t) + l(t)] \tag{6.40}$$

$$\dot{l}(t) = u_n(t) \tag{6.41}$$

where $u_n(t) = \frac{\bar{u}(t)}{M} - \dot{\bar{H}}(y,t)$ is the new control variable.

c. Now let us modify the Lyapunov function by adding the error term to it as in Equation 5.32, which is given by

$$V_a(y,u) = V(y) + \frac{1}{2} l^T(t) l(t)$$

with $l(t) = u(t) - \bar{H}(y,t)$. The control law $u_n(t)$ is given by Equation 5.37 as

$$\begin{aligned} u_n(t) &= -\frac{\partial V}{\partial y} g(y) - kl(t) \\ &= \sum_{i=1}^{6} \frac{\partial V(y)}{\partial y_i} g(y) - kl(t) \\ &= \sum_{i=1}^{6} y_i g(y) - kl(t) \end{aligned} \tag{6.42}$$

This control ensures that the origin of the system (Equations 6.40 and 6.41) is asymptotically stable.

Thus, the final robust feedback control law for Equations 6.28 and 6.29 is given by the following equation:

$$\bar{u}(t) = M[u_n(t) + \frac{\partial \bar{H}}{dy} g(y) u(t)] \tag{6.43}$$

with $u_n(t)$ given by Equation 6.42 and

$$u(t) = \bar{H}(y,t) = H(y,t) - \frac{\eta(y,t)}{1-\kappa(y,t)} w(y,t) \tag{6.44}$$

6.2.2 Robust Control: Matched Uncertainty

Let us consider the uncertainty in the input to the overall system (Equation 6.26) $\bar{u}(t)$. By assuming the uncertainty in input $\bar{u}(t)$ in the second equation we have our input uncertain model as

$$M\dot{u} = \bar{u}(t) + \theta(t) \tag{6.45}$$

$$\dot{y} = g(y)u(t) \tag{6.46}$$

Here the uncertainty is matched as it enters the system at the same point as the actual control input $\bar{u}(t)$. With $|\theta(t,\bar{u},y)| = 0$ the nominal model for this system is given by Equations 6.26 and 6.28. As a first step toward the control design we design feedback control for the nominal model (the system without disturbance) by applying the backstepping technique. After that the robust control for the uncertain system (Equation 6.45) is designed using the Lyapunov redesign technique. We proceed with the control design as follows:

1. **Control by backstepping method:** The first step is to design a feedback controller $\bar{u}(t) = F(y,u)$ for the nominal dynamic system (Equations 6.26 and 6.28) using the backstepping technique. The controller as designed in Section 5.2.4 is given by Equation 5.39 as

$$\bar{u}(t) = M[u_n(t) + \frac{\partial H}{dy} g(y)u(t)] \tag{6.47}$$

with the control $u_n(t)$ given by Equation 5.37 as

$$\begin{aligned} u_n(t) &= -\frac{\partial V}{\partial y} g(y) - kl(t) \\ &= \sum_{i=1}^{6} \frac{\partial V(y)}{\partial y_i} g(y) - kl(t) \\ &= \sum_{i=1}^{6} y_i g(y) - kl(t) \end{aligned} \tag{6.48}$$

where $g(y)$ is the nonlinear system function given by Equation 6.13 and $l(t) = u(t) - H(y,t)$, with $u(t)$ being the control law given by Equation 6.6. The nominal system (Equations 6.26 and 6.28) is asymptotically stable with the control law in Equation 6.47, and there exists a Lyapunov function $V_a(y,t)$ given by Equation 5.32 as

$$\begin{aligned} V_a(y,u) &= V(y) + \frac{1}{2} l^T(t)l(t) \\ &= \frac{1}{2} y^T(t)y(t) + \frac{1}{2} l^T(t)l(t) \end{aligned} \tag{6.49}$$

whose time derivative satisfies the following inequality:

$$\dot{V}_a(y,u) \le -W(y) - kl^T(t)l(t), k > 0. \tag{6.50}$$

2. **Lyapunov redesign:** The next step is to design a robust control for the uncertain model (Equations 6.45 and 6.46) using the Lyapunov redesign method. The goal is to design a feedback control law

$$\hat{u}(t) = \bar{u}(t) + v(t) \tag{6.51}$$

such that we achieve closed-loop stability and asymptotic attenuation of $\theta(t, y, \bar{u})$. $\bar{u}(t)$ is given by Equation 6.47, and $v(t)$ will be designed by using the Lyapunov redesign method. Now with the feedback controller $\hat{u}(t) = \bar{u}(t) + v(t)$ the error in the closed-loop dynamics for the nominal and actual models due to the input uncertain term is

$$\theta(t, y, \bar{u}) = v(t)$$

The magnitude of this uncertain term is given by

$$\|\theta\| = \|v(t)\| \tag{6.52}$$

The bound on the magnitude of uncertainty (Equation 6.52) can be expressed as

$$\|\theta(t, y, u)\| \leq \kappa(y,t)\, \|v(t)\| \tag{6.53}$$

where $\kappa(y,t) > 0$. After applying the control law in Equation 6.51, the rate of change of the Lyapunov function $V_a(t)$ is

$$\dot{V}_a(y,u) = \frac{\partial V}{\partial y} \dot{y}(t) + l^T(t)\dot{l}(t)$$

with

$$l(t) = u(t) - H(y,t)$$

and

$$\dot{l}(t) = \frac{\bar{u}(t)}{M} - \dot{H}(y,t) + \frac{v(t)}{M} + \frac{\theta}{M}$$

Thus, the time derivative of $V_a(t)$ is given as

$$\dot{V}_a(y,u) = \frac{\partial V}{\partial y}[g(y)H(y,t)] + l^T(t)\left[\frac{\bar{u}(t)}{M} - \dot{H}(y,t) + \frac{v(t)}{M} + \frac{\theta}{M}\right]$$

Using the inequality in Equation 6.50, we have

$$\dot{V}_a(y,u) \leq -W(y) - kl^T(t)l(t) + l^T(t)\left[\frac{v(t)}{M} + \frac{\theta}{M}\right] \tag{6.54}$$

By applying the inequality in Equation 6.53 in Equation 6.54, we have

$$\begin{aligned}\dot{V}_a(y,u) &\leq -W(y) - kl^T(t)l(t) + l^T(t)\left[\frac{v(t)}{M} + \frac{\kappa(y,t)v(t)}{M}\right] \\ &\leq -W(y) - kl^T(t)l(t) + l^T(t)\left[\frac{1+\kappa(y,t)}{M}\right]v(t)\end{aligned} \tag{6.55}$$

Let us choose the control law $v(t)$ as

$$v(t) = -\frac{\eta(y,t)}{1-\kappa(y,t)}w(t) \tag{6.56}$$

where $w^T(t) = \frac{l(t)^T}{M}$ and $\eta(y,t) \geq \gamma(y,t)$. Using this control law, Equation 6.55 becomes

$$\dot{V}_a(y,u) \leq -W(y) - kl^T(t)l(t)$$

Hence, we have asymptotic stability with control law (Equation 6.56), As long as $k > 0$. Thus, the robust feedback control is given as $\hat{u}(t) = \bar{u}(t) + v(t)$, with $\bar{u}(t)$ given by Equation 6.47 and $v(t)$ given by Equation 6.56.

6.2.3 Robust Control: Both Matched and Unmatched Uncertainties

Now let us consider both matched and unmatched uncertainties for the dynamic model. This means that there is uncertainty in the input $\bar{u}(t)$ to the overall system (Equation 6.26) and also in the input $u(t)$ to the system in Equation 6.27. Therefore, we have our uncertain model as

$$M\dot{u} = \bar{u}(t) + \theta_1(t) \tag{6.57}$$

$$\dot{y} = g(y)[u(t) + \theta_2(t)] \tag{6.58}$$

Here the uncertainties are both matched and unmatched. The errors θ_1 and θ_2 are given by Equations 6.52 and 6.36 as $\theta_1(t,y,\bar{u}) = v_1(t)$ and $\theta_2(t,y,u) = g(y)v_2(t)$, respectively, and are bounded as $\|\theta_1(t,y,u)\| \leq \kappa(y,t)\,\|v_1(t)\|$ and $\|\theta_2(t,y,u)\| \leq \gamma(y,t) + \kappa(y,t)\|v_2(t)\|$. The control design scheme will be a combination of the designs for matched and unmatched uncertainties obtained in Sections 6.2.1 and 6.2.2. We proceed with the control design as follows:

1. **Control by Lyapunov redesign:** The first step is to design a robust feedback controller $u(t) = H(y,t) + v_2(t)$ for Equation 6.58 using the Lyapunov redesign technique. The control is given by Equation 6.39 as

$$u(t) = H(y,t) - \frac{\eta(y,t)}{1-\kappa(y,t)}w(y,t) \tag{6.59}$$

with $v_2(t) = -\frac{\eta(y,t)}{1-\kappa(y,t)} w(y,t)$ and $H(y,t)$ given by Equation 6.6. The controller stabilizes Equation 6.58 in the presence of uncertainty θ_2. The Lyapunov function is given by Equation 6.33 as $V(y,t) = \frac{1}{2} y^T(t)y(t)$.

2. **Control by backstepping:** The next step is to design a controller for the nominal model (model without uncertainty θ_1) using the backstepping approach. The model given by Equations 6.28 and 6.29 is the nominal model:

$$M\dot{u} = \bar{u} \tag{6.60}$$

$$\dot{y} = g(y)[u(t) + \theta_2(t)] \tag{6.61}$$

The controller $\bar{u}(t)$ is given by Equation 6.43 as the following:

$$\bar{u}(t) = M[u_n(t) + \frac{\partial \bar{H}}{dy} g(y)u(t)] \tag{6.62}$$

The control law $u_n(t)$ is given by Equation 6.42 as

$$\begin{aligned} u_n(t) &= -\frac{\partial V}{\partial y} g(y) - kl(t) \\ &= \sum_{i=1}^{6} \frac{\partial V(y)}{\partial y_i} g(y) - kl(t) \\ &= \sum_{i=1}^{6} y_i g(y) - kl(t) \end{aligned} \tag{6.63}$$

with $l(t) = u(t) - \bar{H}(y,t)$ and $\bar{H}(y,t) = H(y,t) - \frac{\eta(y,t)}{1-\kappa(y,t)} w(y,t)$. The modified Lyapunov function for this system is

$$V_a(y,u) = V(y) + \frac{1}{2} l^T(t)l(t)$$

3. **Control by Lyapunov redesign:** In the final step we design the robust feedback control $\hat{u}(t) = \bar{u}(t) + v_1(t)$ for the overall system (Equation 6.57) using the Lyapunov redesign obtained in Section 6.2.2. The feedback control is thus given as

$$\hat{u}(t) = \bar{u}(t) + v_1(t) = \bar{u}(t) - \frac{\eta(y,t)}{1-\kappa(y,t)} w(t) \tag{6.64}$$

The control law $\overline{u}(t)$ is given by Equation 6.62. The Lyapunov function for this system is the same as in the previous step. Therefore, we achieve asymptotic stability and disturbance attenuation with this controller.

In this chapter we designed robust feedback controllers for kinematic and dynamic models of underwater vehicles. We designed controllers for uncertainty in the input. For the kinematic model the uncertainty is matched and we used the method of Lyapunov redesign, which achieved the attenuation of disturbance. For the dynamic model we discussed both matched and unmatched uncertainties, and a combination of both. We used the method of robust backstepping, which is a combination of Lyapunov redesign and backstepping. In all the controllers we achieved both asymptotic stability and disturbance attenuation.

References

[1] A. D. Luca, G. Oriolo, C. Samson, Feedback control of a nonholonomic car-like robot, in J. P. Laumond (ed.), *Robot motion planning and control*, 170–250, Springer, 1998.

[2] R. W. Brockett, Asymptotic stability and feedback stabilization, in R. W. Brockett, R. S. Millman, H. J. Sussman (eds.), *Differential geometric control theory*, 181–208, Birkhauser, 1993.

[3] C. Samson, Control of chained systems application to path following and time varying point stabilization of mobile robots, *IEEE Transactions on Automatic Control*, 40(1), 64–77, 1995.

[4] J. B. Pomet, Explicit design of time-varying stabilizing control laws for a class of controllable systems without drift, *System and Control Letters*, 18, 147–158, 1992.

[5] A. Astolfi, Discontinuous output feedback control of nonholonomic chained systems, in *Proceedings of the 3rd European Control Conference*, 1995.

[6] C. Canudas de Wit, O. J. Sørdalen, Exponential stabilization of mobile robots with nonholonomic constraints, *IEEE Transactions on Automatic Control*, 37(11), 1791–1797, 1992.

[7] A. R. Teel, R. M. Murray, G. Walsh, Nonholonomic control systems: From steering to stabilization with sinusoids, in *Proceedings of the 31st Conference on Decision and Control*, Tucson, AZ, December 1992, 1603–1609.

[8] A. Tayebi, A. Rachid, A time-varying-based robust control for the parking problem of a wheeled mobile robot, in *Proceedings, of the IEEE International Conference on Robotics and Automation,* April 1996, 3099–3104.

[9] O. Egeland, O. J. Sørdalen, Exponential stabilization of nonholonomic chained systems, *IEEE Transactions on Automatic Control*, 1, 35–49, 1995.

[10] R. M. Murray, S. Sastry, Steering nonholonomic systems in chained form, in *IEEE Proceedings of the 30th Conference on Decision and Control,* December 1991, 1121–1126.

[11] R. M. Murray, S. Sastry, Nonholonomic motion planning: Steering using sinusoids, *IEEE Transactions on Automatic Control*, 38(5), 700–716, 1993.

[12] A. de Luca, G. Oriolo, Modeling and control of nonholonomic mechanical systems, in *Nonholonomic mechanical systems.*

[13] G. Oriolo, A. D. Luca, M. Vendittelli, WMR control via dynamic feedback linearization: Design, implementation and experimental validation, *IEEE Transactions on Control Systems Technology*, 10(6), 835–852, 2002.

[14] A. Isidori, *Nonlinear control systems*, 3rd ed., London: Springer-Verlag, 1995.

[15] O. Egeland, E. Berglund, O. J. Sørdalen, Exponential stabilization of a nonholonomic underwater vehicle with constant desired configuration, in *Proceedings of the 1994 IEEE International Conference on Robotics and Automation*, 1994, 20–26.

[16] M. W. Spong, M. Vidyasagar, *Robot dynamics and control*, New York: John Wiley, 1989.

[17] R. M. Murray, Nilpotent bases for a class of non-integrable distributions with applications to trajectory generation for nonholonomic systems, *Mathematics of Controls, Signals, and Systems,* 7(1), 58–75, 1994.

[18] R. M. Murray, S. Sastry, Steering nonholonomic systems using sinusoids, in *IEEE Proceedings of the 29th Conference on Decision and Control,* December 1990, 2097–2101.

[19] L. G. Bushnell, D. M. Tilbury, S. S. Sastry, Steering three input nonholonomic systems: The fire truck example, in *European Controls Conference*, 1993, 1432–1437.

[20] D. Tilbury, O. Sørdalen, L. Bushnell, S. Sastry, A multi-steering trailer system: Conversion into chained form using dynamic feedback, *IEEE Transactions on Robotics and Automation*, 11(6), 807–818, 1995.

[21] T. Kailath, *Linear systems*, Englewood Cliffs, NJ: Prentice Hall, 1980.

[22] Y. Nakamura, S. Savant, Nonlinear tracking control of autonomous underwater vehicles, in *Proceedings of the 1992 IEEE International Conference on Robotics and Automation,* Nice, France, May 1992, A4–A9.

[23] Y. Nakamura, S. Savant, Nonholonomic motion control of an autonomous underwater vehicle, in *IEEE International Workshop on Intelligent Robots and Systems IROS '91*, Osaka, Japan. November 3–5, 1991, 1254–1259.

[24] G. C. Walsh, L. G. Bushnell, Stabilization of multiple input chained form control systems, in *IEEE Proceedings of the 32nd Conference on Decision and Control*, San Antonio, TX, December 1993, 959–964.

[25] O. J. Sørdalen, M. Dalsmo, O. Egeland, An exponentially convergent law for a nonholonomic underwater vehicle, in *1993 IEEE International Conference on Robotics and Automation,* Atlanta, GA, May 2–7, 1993, 790–795.

[26] H. K. Khalil, *Nonlinear systems*, New York: Macmillan, 1992.

[27] A. J. Healy, D. Lienard, Multivariable sliding mode control for autonomous diving and steering of unmanned underwater vehicles, *IEEE Journal of Oceanic Engineering,* 18(3), 1993, 327–339.

[28] R. Cristi, F. A. Papoulias, A. J. Healy, Adaptive sliding mode control of autonomous underwater vehicles in the dive plane, *IEEE Journal of Oceanic Engineering,* 15(3), 1990, 152–160.

[29] M. J. Corless and G. Leitmann, Continuous state feedback guaranteeing uniform ultimate boundedness for uncertain dynamic systems. *IEEE Transactions on Automatic Control*, 26 (9), 1139–1144, 1981.

[30] L. C. Evans, *Partial differential equations*, Providence, R1: American Mathematical Society, 1998.

[31] A. Isidori, *Nonlinear control systems* II, New York: Springer, 1998.

[32] R. A. Freeman and P. V. Kokotovic, *Robust nonlinear control design: State-space and Lyapunov techniques*, Boston, MA: Birkhauser, 1996.

[33] J. D. Anderson, *Fundamentals of Aerodynamics,* New York: McGraw Hill, 2001.

Index

L

M

N

O

W

Z